Imagining Tomorrow

Imagining Tomorrow

History, Technology, and the American Future

edited by Joseph J. Corn

The MIT Press
Cambridge, Massachusetts
London, England

This book was set in Baskerville by The MIT Press Computergraphics Department and printed and bound by Halliday Lithograph in the United States of America.

Library of Congress Cataloging-in-Publication Data

Main entry under title:

Imagining tomorrow.

Includes bibliographies and index.
1. Technology—Addresses, essays, lectures.
I. Corn, Joseph J.
T20.I43 1986 600 85-15158
ISBN 0-262-03115-9

Contents

Imagining Tomorrow

Introduction

Joseph J. Corn

"The future is not what it used to be," lamented a graffiti artist in a California restroom in the early 1970s. The aphorism voiced a popular sentiment. Anxiety over the human prospect was great, and many commentators were finding little in American life about which to be optimistic. A key element of this alarm was the erosion of faith in science and technology. Millions had begun to question the idea, accepted without reservation just a decade or two earlier, that science and technology inevitably improved the world, that the future would be quantifiably better than the present.[1]

Many kinds of experience informed the growing doubts about technology as an agent of beneficial change and a guarantor of progress.[2] One needed only to look at the environment to realize that rampant technological development was polluting the atmosphere and the waterways, slowly killing forests and oceans as well as animals and plants. Drugs such as thalidomide, which had promised great benefits, instead produced lifelong suffering. The war in Vietnam provided another lesson in the limits of technology, as the most sophisticated weapons systems ever built proved insufficient against a determined enemy. In America's cities, where years of urban renewal and government subsidies had produced modern buildings and transportation systems, there was still no end to economic decline and racial polarization. Finally, as if one needed further reason to be anxious about the future, there were nuclear weapons. The mad accumulation of megatonnage lent credence to visions of global holocaust. A scenario in which humans would cease to exist, a future that promised no tomorrow at all, had become easily imaginable.

It was to this gloomy litany that the graffitist was responding. He was right. By the 1970s the future was not at all what it had once been: sanguine and optimistic. Beginning in the nineteenth century, commentators in industrializing countries had considered social and moral progress to flow inexorably from scientific discoveries and technological innovations. Particularly in the United States, the experience of industrialization and the legacy of evangelical Protestantism had given rise to a view of the future according to which technology seemed to determine progress. For those witnessing the advent of rapid travel by steamboat and the productive miracles of mechanical spinning equipment, it had been hard not to see the future as boundless. The memory of how things had been done before these innovations, along with the tendency to understand secular developments in religious terms, had prompted the celebration of the utopian potential of new machines.[3] A new agricultural implement might be characterized as a "gospel worker,"[4] an agent of transcendent reform and regeneration. From the end of the nineteenth century until the 1960s, it had been common to think of machines as ushering in a better tomorrow, even utopia.

It is the history of the "future that used to be" that is the subject of this book. Because the future exists only in imagination, the authors deal first and foremost with ideas, expectations, and projections. Yet ideas stimulate and justify behavior. Thus, the book is also about those designers, engineers, architects, planners, inventors, and others who acted upon their vision of the future. Although the chapters treat different aspects of this history, collectively they probe the tensions between expectations and developing realities.[5]

All the present authors attended graduate school and came of age intellectually and professionally in the 1970s or the early 1980s. Because we lived through a period in which unalloyed optimism about the benefits of technology was giving way to a more skeptical and anxious viewpoint, our writing and our research came to be suffused by our interest in—and, in some instances, our outrage at—the sanguine expectations that were once widely held.

We found it puzzling that anybody could have seriously expected automobiles to run forever on atomic power, people to commute by private helicopters, or x rays to cure the common cold. How could technology assessment have been so egregiously wrong? How could intelligent people have been so blind to the deleterious impact of technology? The following chapters offer insight into the development of such expectations and their consequences in American history over the last hundred years.

A number of the chapters are case studies of predictions once made about new technologies. For example, Nancy Knight explores lay and medical responses to the discovery of x rays in 1895. Long before

Roentgen's discovery, Knight argues, physicians had tended to believe that eventually they could produce perfect health and everlasting life. But their armamentarium for pursuing that goal was limited. They had nothing comparable to the impressive machines that for over a century had been revolutionizing manufacturing and transportation. Doctors worked on the body with only a few drugs, surgical tools, and the preventive measures of public health. With Roentgen's discovery, however, the medical profession acquired its first complex machine. Physicians and the public envisioned in this device a miraculous potential. In the future, physicians might use rays to look into the nooks and crannies of a diseased body, obtain a quick and reliable diagnosis, and then apply additional doses of rays to effect a cure. Death itself might even be reversible, some prophets suggested. In reality, of course, x rays proved to be a valuable diagnostic and therapeutic tool, yet their curative powers were overestimated and their harmful potential greatly underestimated.

Susan Douglas examines the early history of radio and finds a different relationship between expectation and development. She focuses on the activities of the amateur radio operators (mostly boys and young men) who were active in the years between 1906 and 1912. During that period, Douglas explains, no commercial broadcasting existed, and the government, blind to the potential of radio, did not yet regulate the airwaves. Although officials in Washington failed to glimpse radio's promise, broad segments of the public did so—extravagantly. In vaudeville, newspapers, pulp fiction, and popular-science magazines, enthusiasts promulgated a bold vision of the wireless future. They rhapsodized over how the new technology would permit person-to-person communication without the intervention of telephone companies and their nosy operators. In the wireless age, it was also predicted, life would be less risky, for persons in need would easily be able to summon help. Family members separated by distances too great to be bridged by telephone would be able to talk to one another. Although radio amateurs generally were not the ones making such prophecies, they were in effect already living in the prophesied future. It was they who developed the first broadcasting networks and extolled the idea of freedom of the airwaves. They built their own crystal sets, organized networks of far-flung operators, shared information over the air with unseen colleagues, and relayed distress messages from ships at sea to Coast Guard officials.

In her analysis of the popular enthusiasm for radio, Douglas relies in part on popular-science magazines, a genre of literature that has played an important role in transmitting and shaping expectations for the future. Since the early twentieth century, *Electrical World*, *Science and Invention*, *Popular Science Monthly*, *Mechanix Illustrated*, *Electrical Experimenter*, and *Modern Electronics* have enjoyed broad circulation among boys and young men. These periodicals, which spread information about recent scientific and technical developments and published numerous how-to-do-it and how-it-works articles, also inculcated the philosophy of social progress through technological change in frequent pieces about the future. Articles with titles such as "Miracles You'll See in the Next Fifty Years" contained provocative illustrations of a future world in which ingenious structures, appliances, weapons, vehicles, and other machines would solve virtually every human problem.[6]

The widespread adoption of nuclear power was a favorite topic in popular-science periodicals from the late 1930s into the early 1960s. As Steven Del Sesto notes, however, it was not only the popular-science writers who let their imaginations run riot over the possibilities of nuclear power. Physicists, government officials, and policymakers spoke of the glories of the dawning "atomic age." Particularly after the detonation of nuclear bombs proved the correctness of theoretical assumptions about the power residing in the atom, prophets looked forward to a future with nuclear-powered ships, moon rockets, airplanes, automobiles, and other devices. Enthusiasts spoke of nuclear-generated electricity "too cheap to meter" and expected the mastery of the atom to foster improvements in agriculture, medicine, and even the weather. Del Sesto suggests that some of the men close to the nuclear establishment let political considerations rather than scientific expertise dictate their predictions regarding the peacetime use of the atom and suppressed information about the dangers of radiation in an effort to build support for the government's atomic-energy policy.

While government officials exploited the atomic future to sell policy, various corporations exploited the future to sell products. During the Great Depression of the 1930s, many businesses discovered that allusions to the future in advertising, promotion, and product design could boost sagging sales. The future became associated with streamlined design as manufacturers hired industrial designers who introduced smooth, teardrop-shaped pencil sharpeners, locomotives, adding machines, desks, automobiles, radios, and many other goods that carried associations of efficient and effortless movement. As Jeffrey Meikle has argued elsewhere, the streamlined shape resonated in the imagination of Americans who were anxiously seeking a smooth course through the turbulent Depression.[7] In his chapter here, Meikle explores how a new material, plastic, became associated with the future. Perceived in the early twentieth century as only a substitute for ivory, mother of pearl, and other natural materials, plastic had by the 1920s gained a reputation as the material of the future. By 1940 an author in a popular magazine could opine that the American of tomorrow, "clothed in plastics from head to foot," would "live in a plastics house, drive a plastics auto and fly in a plastics airplane."[8] Behind this shift in opinion lay a growing public respect for science and scientists. In the nineteenth century inventions were popularly perceived as coming from mechanics, but by the twentieth century invention increasingly became the province of scientists. It was chemists who developed plastic. Their transformation of materials and their creation of new ones became symbolic of modern science's contribution to industrial progress, of the promise of the future. Industrial designers also helped make plastic seem the material of the future. They liked it because it was easily fabricated into futuristic streamlined and rounded forms. Corporations whose products could be produced more cheaply with plastics technology also eagerly promoted the new material.

The exploitation of the future to make a profit in the present also constituted a major purpose of corporate participation in the world's fairs of the Depression years. As Folke Kihlstedt demonstrates in his chapter on the Century of Progress Exposition (Chicago, 1933–34) and the New York World's Fair (1939–40), the fairs of the 1930s unabashedly imbued technology with utopian promise. The earliest world's fairs, starting with the World Exposition at London in 1851, had been forums where individual nations and firms exhibited their latest industrial accomplishments. Although the ideology behind those expositions was optimistic and embraced the notion of technology-driven progress, the future was not an explicit theme. At the fairs of the 1930s, however, visions of the future became explicit and utopian. Kihlstedt locates this utopianism in the rhetoric of the fairs' planners and designers, in the architecture of the pavilions, and most notably in the exhibits themselves. He argues that the fairs' planners and designers sometimes borrowed their ideas from earlier utopian visions—most specifically, architectural conceptions and fiction.

Howard Segal provides a larger context for understanding the utopianism so characteristic of much American thinking about the future. Focusing on utopian novels published from 1880 to 1930, Segal finds them to embrace the philosophy of technological utopianism, the belief that machines would do more than politics to bring about a perfect society. Many of the authors he discusses have long been forgotten. One, however, remains well known today and is crucial to an understanding of the history of American technological utopianism: Edward Bellamy. In 1888, Bellamy published *Looking Backward, 2000–1887*, which became an immediate best seller and spawned numerous imitations. Bellamy's young protagonist, Julian West, falls asleep one night in the Boston of 1887 only to awake more than a century later. Through West's recounting of what he sees and experiences, readers learn of the great social and technological changes that are to occur by the year 2000. Nobody works after the age of 44, yet everybody has more than enough food and clothing. The "industrial army" of the future society is so productive that people can charge whatever they need with state-issued credit cards. Because all material wants are amply met, citizens are free to develop their cultural and spiritual interests, aided of course by technology. With their homes connected by telephones to distant locations, they can listen to live musical performances or church services without leaving their living rooms. Outside their homes, the once-dingy industrial city has become a place of beauty, its skyline characterized by a "complete absence of chimneys and their smoke."

Since Bellamy, the home and the city of the future have been the concerns of many American visionaries, and these are the subjects of two of the chapters here. Brian Horrigan traces the changing appearance and meaning of the "home of tomorrow" concept from the late 1920s through World War II. He demonstrates how style, economic conditions, and technical developments all affected popular images of the home of tomorrow. Furthermore, he documents the way business has exploited and even manufactured the future as a marketing tactic. For example, in 1934 Westinghouse built a prototype home that utilized the electrical power of thirty average homes to operate a veritable laboratoryful of electrical devices. Ultimately, Horrigan shows us, the Westinghouse vision of the home of tomorrow as a "container for appliances" won out over two competing conceptions: the "machine for living," as the Swiss avant-garde architect Le Corbusier described the modern house, and the home as a standardized mass-produced product that would roll off assembly lines like Fords.

If people were lukewarm to the idea of standardized and systematically planned homes, they felt differently about cities. As Carol Willis argues in her chapter on the city of the future, many Americans in the 1920s and the 1930s embraced the vision of a uniform, planned, skyscraper-dominated city. This was a change from the negative view of all things urban that had colored American thinking since Thomas Jefferson and had caused most commentators to view turn-of-the-century Chicago and New York, dominated by their new skyscrapers, as overcrowded, congested, and hectic zones of social decay and predatory capitalism. In the 1920s, however, led by a new vision, popular thinking about the city became more positive. Professional city planners and architects began to articulate an image of a future city that would be as clean, efficient, ordered, and beautiful as existing cities were dirty, chaotic, haphazard, and ugly. Harvey Wiley Corbett, Raymond Hood, Robert Lafferty, and Francisco Mujica drew up plans to recast the form and appearance of the American city. Their visions of the future city became widely known as renderings were published and models exhibited. Similar cities of tomorrow dazzled millions in futuristic films such as *Just Imagine* and in exhibits such as General Motors' Futurama at the New York World's Fair. For the first time in American history, Willis concludes, the city—in its imagined futuristic incarnation—became an accepted site for utopia.

The visions of the future city discussed by Willis came closest to fruition in the United States in the period of urban renewal after World War II, when federal subsidies encouraged the bulldozing of decaying districts and the construction of towering apartment houses and other structures. Yet the result was anything but utopia. Cold, forbidding, and lacking in human scale, many of the buildings put forth in the name of urban renewal in the 1950s and the 1960s gave a bad name to the architects whose earlier visions they echoed. The experience demonstrated, however, that the ideas and expectations people hold about the future inevitably have historical consequences, for better or for worse.

Sometimes a particular vision of the future is influential in the process of technological development. Paul Ceruzzi demonstrates this in his chapter on the "unforeseen revolution" in computers. He begins with the fact—sobering to recall for those who believe that technological forecasting or "technology assessment" might become a true science[8]—that in the 1940s computer pioneers commonly assumed that a half-dozen or so computers would adequately serve the industrial world for the foreseeable future. To explain this dazzling failure of prophecy, Ceruzzi shows how the backgrounds of the computer pioneers (most of whom were physicists) and the experimental traditions of their discipline shaped expectations for the new machines. These men had often built large and costly pieces of experimental equipment, used them a few times, and then relegated them to the scrapheap or the storeroom. They tended to view the early computer as one more experimental device, not as the first of a whole new breed of general-purpose machines. Another reason for their inability to foresee a computer revolution was the fact that the calculations performed on early computers were quite exotic, which made it hard to imagine that such machines might be applied to wholly different kinds of work.

Experience, then, not simply technical constraints, Ceruzzi argues, underlay the inventors' failure to envision a future for the computer anything like that which actually unfolded. In turn, their limited expectations influenced the early commercial development of the computer. Prophecy, as it were, became self-fulfilling. Perhaps the computer revolution now upon us might have started earlier had the pioneers held different expectations of their invention.

Carolyn Marvin's chapter on the electric light examines another way in which ideas about the future intertwine with and influence behavior. In the late nineteenth century and the early twentieth century, the electric light was widely perceived as a potential medium of communication, not simply as an illuminating device, and this view motivated many of its uses. For instance, light bulbs appeared in stage actresses' costumes and as theatrical props. "Sky signs" containing hundreds of bulbs flashed news bulletins, celebratory wishes, or advertisements from the tops of tall buildings. Advertisers experimented with the use of searchlights to project messages onto clouds. Some enthusiasts even speculated about shining beams on the moon or Mars. By examining these early applications of electric lighting (actual and projected)—a foreshadowing, Marvin suggests, of the electronic mass media—it becomes clear that predictions about the future have subtle and sometimes unexpected consequences.

I consider the relationship between predictions and consequences at greater length in the epilogue, where I also reflect more generally upon the phenomenon of prediction in history. Clearly, the vision of the future as a technological paradise has been a central theme in American culture. This will become evident as we turn now to examine the future that "used to be."

I extend my thanks to the authors for their creative responses to criticism and their willingness to rewrite more than they might have wished. Finally, to Wanda M. Corn, T. J. Davis, Brian Horrigan, and William Rorabaugh: I am deeply appreciative of your friendship and your excellent editorial suggestions.

Notes

1. The graffito is quoted in Michael Davie's *In the Future Now: A Report from California* (London: Hamish Hamilton, 1972), p. 225. A brief review of some of the "antitechnology" literature and its emergence in the 1960s and the early 1970s, written by one of technology's most eloquent defenders, can be found in chapter 4 of Samuel C. Florman's *The Existential Pleasures of Engineering* (New York: St. Martin's, 1976).

2. Symptomatic of the shift in public sentiment regarding technology's promise were the various political movements directed against particular technological applications,

such as the successful effort to prevent the development of an American supersonic transport. See Mel Horwitch, *Clipped Wings: The American SST Conflict* (Cambridge, Mass.: MIT Press, 1982).

3. See "Improved Hay Maker," *Scientific American*, n.s., 2 (March 1860), p. 216, quoted in Leo Marx, *The Machine in the Garden: Technology and the Pastoral Ideal in America* (New York: Oxford University Press, 1964), p. 198.

4. John C. Kimball, "Machinery as a Gospel Worker," *Christian Examiner* 87 (November 1869), p. 319. On general attitudes toward technology in nineteenth-century America see Marx, *The Machine in the Garden*; John Kasson, *Civilizing the Machine: Technology and Republican Values in America, 1776–1900* (New York: Viking, 1976); Thomas Parke Hughes (ed.), *Changing Attitudes Toward American Technology* (New York: Harper and Row, 1975).

5. On the history of the future, see I. F. Clarke, *The Pattern of Expectation: 1644–2001* (New York: Basic Books, 1979). For a comprehensive study of predictions about technology in the future, see George Wise, Predictions of the Future of Technology: 1890–1940, Ph.D. diss., Boston University, 1976. See also Kenneth M. Roemer, *The Obsolete Necessity: America in Utopian Writings, 1888–1900* (Kent, Ohio: Kent State University Press, 1976). A number of studies have focused on the expectations Americans have projected onto particular technologies; see, e.g., Vary T. Coates and Bernard Finn, *A Retrospective Technology Assessment: Submarine Telegraphy, The Transatlantic Cable of 1866* (San Francisco Press, 1979), especially chapter 4; James W. Carey and John J. Quirk, "The Mythos of the Electronic Revolution," *American Scholar* 39, no. 2 (1970), pp. 219–241; 39, no. 3 (1970), pp. 395–424; Joseph J. Corn, *The Winged Gospel: America's Romance with Aviation, 1900–1950* (New York: Oxford University Press, 1983).

6. Waldemar Kaempffert, "Miracles You'll See in the Next Fifty Years," *Popular Mechanics*, February 1950, pp. 112–118, 264, 266, 270, 272. Little has been written about popular-science magazines, especially in the twentieth century, but see Matthew D. Whalen and Mary F. Tobin, "Periodicals and the Popularization of Science in America, 1860–1910," *Journal of American Culture* 3 (spring 1980), pp. 195–203. On Hugo Gernsback's science-fiction publications, but not his more factual popular-science magazine *Science and Invention*, see Paul A. Carter, *The Creation of Tomorrow: Fifty Years of Magazine Science Fiction* (New York: Columbia University Press, 1977), passim.

7. Jeffrey L. Meikle, *Twentieth Century Limited: Industrial Design in America, 1925–1939* (Philadelphia: Temple University Press, 1979).

8. Julian P. Leggett, "The Era of Plastics," *Popular Mechanics* 73 (May 1940), p. 658, quoted by Meikle in chapter 4 of this volume.

9. The literature on technology assessment is vast and growing, but it is oriented toward the future rather than the past. A number of historians, however, are conducting what they call retrospective technology assessments in the belief that the past has something to teach us about the prospective effects of present technology. See Joel A. Tarr (ed.), *Retrospective Technology Assessment—1976* (San Francisco Press, 1977); Coates and Finn, *A Restrospective Technology Assessment*. A useful introduction to this literature is Howard P. Segal's "Assessing Retrospective Technology Assessment: A Review of the Literature," *Technology in Society* 4 (1982), pp. 231–246.

1

"The New Light": X Rays and Medical Futurism

Nancy Knight

Planners and dreamers since the time of Plato have posited ideal future worlds in which all diseases are curable and lives are long and happy.[1] The two most constant images in these felicitous projections are the biologically perfect life and the unlimited life span. From early legends of the miraculous "fountain of youth" to the Victorian ideal of a sanitized city combining "the lowest possible general mortality with the highest possible individual longevity,"[2] an idealized eternal life has remained the implicit goal of medical progress. Earlier dreamers would have found the bald statement of such a goal irrational and blasphemous, flying in the face of a terrestrial predestination to contagious disease and premature death from any number of seemingly foreordained causes. In the last century, scientists have hesitated to state flatly the goal of biologically perfect eternal life, but it has been implied in endeavors as diverse as the elimination of contagious disease, eugenic intervention, and genetic engineering.

A perfect life that lasts forever has been a constant human dream, but the medical routes envisioned to achieve it have been varied. Thomas More, in his desire to construct an ideal political society for sixteenth-century England, gave scant attention to science and medicine. Of medicine, his narrator commented simply: "yt ts a wyse mans parte rather to avoyde sycknes, then to wyshe for medycynes."[3] Here More neatly summarized the two directions medical futurism would take in the following four centuries: a striving for cooperative prevention and avoidance of disease, and recurrent hopes for miraculous discoveries that would spontaneously eliminate illness and prolong life. These directions are not antithetical; they coexist in many projections about future health. Yet a subtle tension exists between the notion of perfect health achieved by cooperation, prevention, and research and the more exciting promise of instant miracle cures. At various times, each direction has had its turn in holding the balance of speculative attention.

I will focus here on the shift in medical futurism in the years surrounding the discovery of x rays in 1895. For much of the nineteenth century, the focus of popular and professional medical prognostication was on the prevention of socially aggravated maladies. With the discovery of x rays, the basis for the first "miracle machine" in clinical medicine, a change occurred in medical dreams. It seemed to many that machines might transcend traditional healing powers and that solutions to disease and death were as close as the nearest patent office.

Shifts in ideas about the medical future have reflected contemporary concerns. In the nineteenth century, industrialization and urbanization gave rise to a scientific point of view that increasingly prized structure and organization. Physicians and laypersons began to speculate on a medical future shaped by highly structured efforts in the areas of public health and sanitation. This tradition reached its culmination in the extraordinarily influential work of the Englishman Benjamin Ward Richardson, whose *Hygeia: A City of Health* (1875) was the ultimate exercise in sanitary logic and the origin of suburban planning in Great Britain. In Hygeia, a precisely controlled garden city of 100,000 persons living in 20,000 houses on 4,000 acres, the "perfection of sanitary results" would be achieved through scientific planning and cooperation.[4] Richardson and other hygienists had little time for what they viewed as useless speculation about easy solutions to complex social ills.

The most influential American future-thinker of the nineteenth century, Edward Bellamy, gave scant attention to the technological possibilities for medicine. Bellamy's *Looking Backward* (1888) shows little evidence that he was influenced by the introduction of new vaccines or by such a medical breakthrough as the use of antisepsis (cleanliness measures to fight disease-spreading microorganisms). His narrator is much more concerned with the education and finances of physicians in the year 2000 than with the details of their practice. Medicine is socialized, and the prevalence of good health is due more to organizational achievements than to science. The life span has indeed increased, the narrator notes, but this is due to "the better conditions of existence nowadays, and above all the freedom of everyone from care."[5] Even the "women's diseases" of Bellamy's time are cured in his perfect future, not by nostrums and miracle medicines, but by furnishing each woman with a "healthful and inspiriting occupation."[6]

Most of the legion of American futurists who immediately followed Bellamy were also more concerned with organization than with the actual mechanics of medical practice. Many failed to mention medicine at all. Those who did, such as the carriagemaker Chauncey Thomas in his book *The Crystal Button: Or, Adventures of Paul Prognosis in the Forty-Ninth Century* (1891), used futuristic medicine as a device for contemporary criticism. Thomas's narrator in the year 4872 notes that most diseases of the late nineteenth century were "mainly attributable to poor or inappropriate food, poor drainage, lack of sunshine and fresh air, lack of exercise or too much of it, vice of many kinds, and ignorance of even the simplest laws of physical well-being."[7]

In the early 1890s, even those projections that were aimed specifically at elucidating the possible future of the age's new industry and technology saw few medical miracles in store. Observers of medicine looked for the equivalents in their field of the bicycle and the Ferris wheel (or, more serious, those of the telegraph and the electric light), but found none.[8] Some scientists seemed to think that medical knowledge was complete and needed only to be refined with time. Americans had already made the great discoveries in all areas of science, according to an 1894 study entitled "National Progress and Character," and future generations would merely fill in the gaps in the nineteenth century's knowledge and accomplishment.[9] In the same year, a writer for the Smithsonian Institution saw great promise in air travel and improved marine navigation, but no great changes in medicine and biology. The only future health concern he elaborated was a hope that animal fodder might be made suitable for human consumption where other comestibles were scarce.[10]

To some observers, however, the climate of American industry and technology in the last decade of the nineteenth century was alive with possibilities—especially in the somewhat unexciting areas of orderly prevention and sanitation. In a rapidly industrializing society in which wonders of modern technology such as the light bulb were changing the average home, it was not inappropriate for a few observers to expect something grand from medicine. Electrotherapeutics—the treatment of a variety of ailments by the surface application of static electricity—seemed to embody the spirit of the age. In a piece detailing electrotherapeutic exhibits at the 1893 Columbian Exposition, one author noted that "now nearly every well-informed man appreciates the probability of almost miraculous progress in the immediate future" of electricity in medicine.[11] But the medical future, even for optimists like the electrotherapists and the electricians, remained unclear. While some futurists maintained that sanitation and the organization of extant techniques and knowledge would characterize the future of health, others dreamed of great breakthroughs but could not articulate the content or form of these medical miracles. These optimists would not have to wait long for inspiration. It would arrive, dramatically enough, in the form of a "new kind of light."[12]

The event that changed the shape of the future for many medical observers occurred on November 8, 1895, in Germany. The physicist Wilhelm Conrad Roentgen had been experimenting with the production of cathode rays using a common piece of laboratory equipment, the Crookes or cathode tube. Cathode rays, first investigated widely in the 1870s, were used by physicists in experiments to determine the true nature of light. Working in a darkened room, Roentgen charged his tube with electricity and was soon distracted by a green fluorescence issuing from metals and other materials on his work table. Especially bright was a barium-platinocyanide crystal screen of the type used by physicists to detect cathode rays outside vacuum tubes. Roentgen knew that, since the fluorescence extended to objects 7–8 feet away from the light source, he could not be observing the familiar short-distance cathode rays. After only a little experimentation, he discovered that placing a hand or arm between the charged tube and a fluorescent plate or screen rendered the bones clearly visible through the flesh. In a few days he produced a photograph of his wife's hand, with rings and bracelet. By January 1, 1896, Roentgen had circulated reports of his discovery of "A New Kind of Ray" throughout Europe and the United States.[13] The central theme of twentieth-century medical technology—the total visibility of living human organisms—had been introduced.[14]

In the first months after Roentgen's announcement, the medical profession and the public were treated to predictions of immediate miracles and potential uses for the "new photography" or the "new light." The editors of *Scientific American* were at first skeptical of the x-ray technique, but in February 1896 they summarized recent changes in public attitudes: "It seemed as if the limits of human discovery were being reached, but the wonder of the new photography only emphasizes the possibility of other victories to be won in the world of science."[15] They believed that the new discovery was so revolutionary that it "almost dangerously increases our powers of belief."[16] For both the professionals and the public, these powers of belief were often severely strained; however, it is clear that, whether the resulting product was humorous or serious, the x ray had touched the popular and scientific imaginations.

People reacted to the discovery of the x ray in three general ways, each relating differently to medical futurism. First, there was an immediate popular response that spawned the sort of cultural manifestations common to fads. X rays appeared in advertising, songs, and cartoons. Amidst all this there was a popular rush to speculate on what the new technology might achieve in the near future. A second reaction occurred within the medical and scientific communities, where response was equally excited but was generally tempered by a more practical awareness of immediate possibilities. Physicians and scientists speculated on the various ways in which x rays might do everything from illuminate the interior of the human brain to explain the nature of light. Yet most of these speculations contained a kernel of scientific fact or plausibility. The most interesting aspect for the scientific community in this sense was the choice of experiments with x rays. All scientific experiments have an element of "futurism": the investigation of a hypothesis about an unknown in the present that may be somehow useful or fruitful in the future. Certainly there was something of a medical dreamer in the physicist who attempted to rejuvenate dead mice with an x-ray apparatus.[17] Many other early applications of x-ray machines show the extreme hopes for health and longevity that were vested in them.

X rays entered the collective imagination through novels and short stories about the future. More coherent than the early x-ray-craze predictions and more extravagant than the statements of physicians and scientists, this literature demonstrated the range of popular interest in the "new light." And it was in fiction, where speculation was at its freest, that the change from a cooperative view of future public health to one of "miracle" machines and panaceas was most apparent.

The x-ray mania began early and grew quickly. "Hidden Solids Revealed!" trumpeted the *New York Times* in January 1896.[18] The press was enchanted with the possibilities of the new rays. With the information that they rendered "Wood and Flesh More Easily Penetrated . . . Than Plain Glass,"[19] many observers immediately speculated on various applications and uses. Even the most mundane experiments with the new technique were labeled miraculous. "Startling results" announced by professors at Yale turned out to be x-ray photographs of uncracked walnuts showing "a splendid view of the kernels."[20] Some popular magazines and journals showed x-ray photographs of feet in boots, coins in wooden boxes, and shapely women in tight lacing. One popular cartoon (figure 3) hinted at the possible leveling effects of the rays by revealing that beneath the superficial layer the well-to-do of the Gilded Age were the same as the common people.[21]

Because of the simplicity of x-ray apparatus, access to the new technology was soon available to all interested parties, amateur or professional. This widespread availability seems to have fueled the

THE MARCH OF SCIENCE.

INTERESTING RESULT ATTAINED, WITH AID OF RÖNTGEN RAYS, BY A FIRST-FLOOR LODGER WHEN PHOTOGRAPHING HIS SITTING-ROOM DOOR.

Figure 1
Cartoon, *Punch* 110 (1896), p. 117.

THE NEW ROENTGEN PHOTOGRAPHY.
"LOOK PLEASANT, PLEASE."

Figure 2
Cartoon, *Life* 27 (February 27, 1896), p. 155.

notion that x rays would soon be a part of everyday life. All that was needed to produce them was a Crookes tube (an evacuated glass tube with an anode and a cathode) and a power source. Reliable electrical service with predictable, stable voltage was not widely available at the time, but this did not hinder x-ray enthusiasts. The new machines were powered by induction coils, which could be obtained from any one of a growing number of electrical machine shops, or by static generators, which were a ubiquitous staple of the electrotherapeutic movement in medicine. Tubes were available in most laboratories and could easily be reproduced in quantity by any glassblower. Simple glass photographic plates coated with a fluorescent emulsion completed the early x-ray outfits. In short, almost anyone with a little scientific knowledge in these areas could enter the field, and it seemed that in the early days almost everyone wanted to experiment with x rays.

WHETHER STOUT OR THIN, THE X-RAY MAKES THE WHOLE WORLD KIN.

Figure 3
Cartoon, *Judge* 32 (1897).

Speculation on the possibilities of x rays ranged from the mundane to the fabulous. An Idaho man claimed that his own invention, the "cathoscope," would enable chicken farmers to distinguish fertilized from unfertilized "hen fruit."[22] The household possibilities of x rays were extolled by a woman who lost her rings while baking a cake. Not wanting to ruin the cake, she took it to a photographer, who pinpointed the rings with x rays for easy removal.[23] The possibly apocryphal individual who mailed Thomas Edison a pair of hollow binoculars and asked to have them fitted with x rays was one of many who believed that the new technique would extend beyond the laboratory and the clinic.[24]

Lead underwear to protect ladies against x-ray-equipped peeping Toms was advertised.[25] A Henry Slater advertised in 1896 that, owing to the success he had already achieved with "the New Photography," he was "prepared to introduce same in divorce matters free of charge."[26] A popular stage illusion purported to "x-ray" a bedroom so that the audience could see through the walls to what happened within.[27]

Although the editors of *Electrical Engineer* believed that "very few people would *care* to sit for a portrait which would show only the bones and the rings on the fingers," others saw promise in the future of skeletal photography.[28] Not the least element in the attraction to the x ray was the mixture of fear and awe inspired by the photographs. One observer called the public reaction one of "wonder not altogether unassociated with a certain horror."[29] Nevertheless, people were lining up for 30-minute or one-hour sittings for a chance to see their bones.[30]

Some observers envisioned educational and socially constructive applications of the rays. One odd but popular invention was the "x-ray slot machine." In some Manhattan and Chicago restaurants, a nickel deposited in such a machine sent electricity through a fluoroscope, allowing the moving bones of the depositor's hand and wrist to be seen. This demonstration was advertised as educational, "calculated to give the man in the street a glimpse of natural phenomena that he might not otherwise obtain."[31] Inventive reformers found in x rays the future solution to problems of alcohol abuse: By showing a drunkard a photo revealing the ravages of alcohol within his body, one might get him to sign the pledge immediately.[32] (A widely circulated cartoon made light of this effort, intimating that the skeletons of alcoholics were revealed by x rays to be loosely hinged structures of whiskey bottles and shot glasses.[33]) Some physicians varied the instructive approach of the reformers, seeing the potential of using x rays to convince complaining patients that they had no actual organic problems.[34]

X rays, many believed, would become a part of everyday culture, from henhouses to the temperance movement, from the detection of flaws in metal to the analysis of broken hearts. One indication of this widespread belief was the number of humorous verses about the rays. A typical example:

Not worth your while
That false, sweet smile,
Which o'er your features plays:

Thy heart of steel
I can reveal
By my cathodic rays![35]

Some believed seriously that x rays would provide access to thoughts and feelings. The discovery of x rays coincided with a rise in public interest in psychic and supernatural phenomena, and soon the new technology and the old preoccupation were linked. Very early in the excitement over x rays, an Italian physician was widely quoted as suggesting that spiritualists who claimed to see through opaque matter might have retinas "sensitive to the x-rays."[36] The notion of x rays

Figure 4
The apparatus used for one of the first public demonstrations of x rays, Chicago, February 8, 1896. From *Western Electrician* 18 (1896), p. 73.

THE FRENCH X-RAY CUSTOM-HOUSE GLASS MAY BE JUST THE THING APPLIED DURING THE COMING SEASON.

Figure 5
Cartoon, *Life* 30 (1897), p. 354.

as a sensory extender was reinforced by numerous confirmations that blind persons could "see" the rays, or that radiation caused a lasting fluorescence to be visible to the blind.

The most famous of the early speculations about extrasensory use of x rays came in September 1896 in *Appleton's Popular Science Monthly*. In his article "The Sympsychograph: A Study in Impressionist Physics," David Starr Jordan gave detailed information on the development of a machine that could purportedly photograph thoughts."[37] Jordan, then president of Stanford University, probably intended the article as a joke, but *Appleton's* carried it in all seriousness, and it was widely reprinted. According to this article, after inspiration from x rays, Jordan and his colleagues discovered that "just as one sensitive mind at a distance receives an image sent out from the psychic retina of another, so could the same magic be concentrated and fixed upon a photographic plate."[38] The sympsychograph was an electrical camera with a faceted lens, with each facet connected by tubes and wires to the eyes of the participants. Each person was told to concentrate his thoughts upon a cat, and the resulting photo was described generously by *Electrical World* as "a number of indistinct images of a cat."[39] In fact, the item was a blurred composite of seven photographs. Jordan later published a retraction, but many people continued to believe that x rays would provide a link with the unknown.

From the examination of chicken eggs to communication with another dimension, hopes for the new technology reflected a wide spectrum of contemporary concerns. Even old alchemical dreams were revived; a man in Cedar Rapids claimed to have changed "a cheap piece of metal worth about 13¢ to $153 worth of gold" by the skillful application of x rays.[40] Others saw in the cathode tube the "promise of artificial light of the future."[41]

A second kind of response to the rays came from the medical community. While the press engaged in various nonmedical speculations, physicians claimed to see a future role for radiology in every aspect of life, from birth to death and beyond. The public learned that x rays might soon be used routinely for everything from diagnosing pregnancy to raising the dead. Not only did medical observers speculate about the diagnostic and prophylactic possibilities of the rays, many hoped that they would cure a wide variety of diseases. Some even suggested that x rays would solve the riddle of cancer and provide important clues to extended longevity.

Nothing could have seemed so fantastic to medical diagnosticians as the news that one could now see through the skin and bones of a patient and instantly divine the nature of an illness or an injury. Few physicians doubted the power and efficacy of x rays. A few expressed skepticism about the range of predictions, however. A year and a half after the discovery, one medical observer scoffed: "With no discovery within my recollection has the immediate and general excitement been so intense, or the subsequent deluge of absurd and improbable statements so great as with this discovery of Prof. Roentgen."[42] It must have been extremely difficult to distinguish the absurd and improbable from the realistic and verifiable. To modern eyes, almost all the early experimental applications of x rays seem to have had elements of daring and hope. The astonished physician who in early 1896 wrote "*Imagine*, that when this discovery has been brought to perfection, we shall be able to detect foreign substances in any part of the body" was considered no less extravagant than the man who proposed to resuscitate the dead with x rays.[43]

The most obvious hope for x rays was that they could be used to diagnose illnesses and to observe the living body. Much of the early speculation on the diagnostic application of the rays was merely the logical extension of their documented powers. In April 1896, a doctor in Glasgow stated that he did not think it a dream that x rays would eventually reveal all the body's functions. He noted that "in the surgery of the future the surgeon going his daily rounds would be able to look through splints and bandages and see the condition of fractures."[44] Such a prediction might have seemed farfetched to contemporary medical workers; nevertheless it was a rational extension of the already demonstrated powers of x rays.

Until well into the twentieth century, all other causes for symptoms had to be ruled out before a woman was considered pregnant. X rays seemed to provide a perfect solution to this problem. Although early trials were attempted on cadavers, soon live fetuses were being studied *in utero* with the technique. One woman, 8 1/2 months pregnant, submitted to an x-ray exposure of 75 minutes in order to secure a clear picture of the fetus.[45] Not until the danger of burns and the possibility of induced carcinoma were well established, in 1903–04, did physicians give up hope that x rays might become the standard tool for affirming a diagnosis of pregnancy.[46] The principle of pediatric diagnosis was, however, combined with the role of the soothsayer by the physician who believed that fluoroscopy of the young could be used to "determine their later susceptibilities."[47] (Children might be x-rayed, he believed, to discover latent diseases to which they might later be prone.)

Thomas Edison, with his usual enthusiasm for new technologies, believed that the "new vision" would allow inspection of the body's innermost recesses. On February 8, 1896, he announced that he was on the verge of making an x-ray photograph of the human brain at work (besides attempting to "find some practical commercial use" for x rays).[48] The papers reported daily on Edison's attempts to focus enough radiation to penetrate the bony structures of the human skull. By mid February he announced a delay, and, although he never succeeded, he firmly believed that x rays would someday reveal the functioning of the human brain.

Much more sensational than the diagnostic and exploratory applications of x rays were the many therapeutic possibilities envisioned. One of the most promising of these was that the rays might cure or correct blindness. One medical author wanted to use them to "imprint visual images" on the retinas of blind persons, enabling them to recognize documents, individuals, and work-related items.[49] Another physician tried focusing x rays on the eyes of the blind; although his results were inconclusive, he believed that he had hit on an interesting clue to the nature of blindness.[50] Throughout the late 1890s and the early 1900s, short reports on the use of the x rays to heal blindness and restore sight appeared in medical journals. No doubt this interest was spurred by the lingering fluorescence often detected by the blind after exposure to radiation.

The medical futurism of any age is grounded in immediate contemporary concerns. One focus of medical attention at the turn of the century was the rather recent discovery that germs and bacteria cause disease. Of special interest in the 1890s was the tuberculosis bacillus. With no real basis in experimental fact, x rays were seized upon by the medical and scientific communities as a potential wide-spectrum bacteriocide and as a cure for the contagious diseases caused by bacteria—especially tuberculosis. It was hoped that x rays would supersede the complex sanitary and prophylactic precautions of preventive medicine. As early as February 1896, Edison asked the public "What can be easier than to turn the rays on the lungs of persons afflicted with consumption?"[51] Soon his theory that x rays could kill bacilli was being tested in laboratories and clinical settings throughout the United States and Europe. Popular journals speculated that diphtheria and tuberculosis might easily be checked by "exposure of the germs to the rays, either directly or through the flesh."[52] Undoubtedly drawing on the popular "Finsen red light treatment," first used to treat smallpox in the early 1890s, experimenters soon trained x rays on various forms of bacteria. Results were not encouraging, but the experimenters were undaunted in their enthusiasm. Many researchers reported positive effects from the application of the rays to tuberculosis of the lungs.[53] Both lupus vulgaris, with its characteristic tubercular skin lesions, and lupus erythematosis, then believed to be of bacterial origin, showed remarkable responses to x-ray exposures, and this encouraged more speculation on the therapeutic future of the technology.[54] Some researchers believed they had proved conclusively that tuberculosis bacilli and staphylococcus cultures could be destroyed both by exposure to x rays in laboratory dishes and by direct exposures of the patient.[55] Others reported good results in the laboratory destruction of the cholera spirillum.[56] The most common explanation for the so-called bacteriocidal action of the rays was that they surrounded bacteria with "a medium which destroys or dilutes them." This explanation was soon to be supported by the observation that x rays also had extremely adverse effects on human tissues.[57]

In fact, the most extravagant and ultimately the most fruitful experimentation with x rays was due to the fact that, in sufficient strength and duration, they destroyed almost everything in their path. It may have remained doubtful that the rays could destroy bacteria, but it was a proven fact that prolonged exposure of patients and doctors caused their hair to fall out. Early on, this was viewed as a potential benefit of the technology: "Should it be ascertained that the hair does not grow again, we should have a very simple method of depilation."[58] In France, x-ray clinics were set up for the removal of facial hair in women, and American writers continued to believe that x rays could be a valuable cosmetic aid, "especially in cases where the growth is diffuse and profuse."[59] The destructive possibilities of x rays were often carried to extremes; one physician used the rays to painlessly kill laboratory guinea pigs for experimentation, and cheerfully suggested that "there is a chance for useful original work in this field" of destructive x-ray energy.[60]

Found to be effective against a wide range of skin diseases and lesions, x rays were soon turned on cancers. Around 1895, articles alleging an increase in cases of all types of cancer began to appear in American medical journals.[61] Using new statistical techniques for evaluating epidemiological reports, doctors decided that cancer was increasing at an alarming rate. Whatever the validity of their conclusions, it is true that reported cases of cancer were on a sharp increase at the close of the century and that physicians were anxious over their impotence when faced with most carcinomas. Mixing hope, truth, and fiction, one doctor summarized the general attitude toward the use of x rays against cancer:

> I must confess that, as with every new discovery, my thoughts turn to the possible application to a subject which has an invincible attraction for me—the cure of cancer. I was led in a purely speculative and empirical manner to direct the use of the X-rays in a number of inoperable cases of cancer in various regions. My small hopes of success were bolstered up somewhat by reflections upon the hypothetical bacterial cause of cancer, upon the wonderful effect of sunlight on the tubercle bacillus, upon the remembrance that the most powerful germicidal effects of the sun are caused by the rays near the violet end of the spectrum and having the most chemical activity . . . , and that Roentgen rays are ultraviolet and possess chemical activity.[62]

Others did not construct such elaborate excuses for using x rays on cancer; they plunged ahead with no apparent rationale other than that the technique was new and untried. As early as January 12, 1896, less than two weeks after Roentgen's announcement, an electrotherapist in Chicago, Dr. Emil Grubbe, exposed a woman's cancerous breast to x rays.[63] It is not clear how successful Grubbe's treatment was, but soon many forms of both surface and deep-seated lesions were being treated by prolonged exposures. Widespread reports of cures and successes, as well as the unusual pain-relieving effects of the rays noted by many cancer patients, raised hopes and heightened speculation.[64] Not only were x rays deemed potentially useful in isolated cancers, but soon they were applied to deeper problems. Good results were obtained in the x-ray therapy of several types of leukemia.[65] Reports of interior cancers cured through apparently healthy tissue gave even more hope.[66]

Many physicians wondered whether x rays might be used in novel ways and in combination with other techniques to cure cancer and to prevent its spread. In 1903 one visionary proposed that "in the future . . . in a case of cancer of the breast, with involvement of the axillary glands, and some evidence of extension to the bronchial structures, I think it would be the proper thing to remove the great mass of the disease by surgical operation, and then apply the X-rays until all evidence of the disease disappears."[67] Another physician wondered if it might "be possible in some way to extend the action of this agent [x rays] . . . to obtain the destructive effects on cancer at all depths," perhaps "by means of some chemical agent introduced either into the general circulation or locally injected."[68] Espousing his own view that light is composed of particles rather than waves, the physicist Nikola Tesla theorized that it might be possible to "load" x rays with cancer-fighting drugs or chemicals and "project" these therapeutic substances into specific body parts.[69]

If x rays promised hope for cancer patients, they might even prevent death itself; so suggested some of the more extravagant claimants. Just a month after the discovery of x rays, a professor at the City College of New York drowned a mouse for later experimentation with the rays. (Live animals would not hold still for long exposures.) On the application of the rays, after 10 minutes' submersion, the mouse revived. Astonished, the professor began experimenting with other mice and with garter snakes "to find out if either the Roentgen rays or the ozone produced so plentifully by the electrical apparatus could possibly be concerned in the apparent restoration to life."[70] The results were negative, but the possibility of using x rays to prolong life was raised again by no less a personage than Marie Curie. In an experiment on the effects of radium on mealworm larvae, she noted that most died, but some stayed alive in the larval stage long after the control group had grown old and died. Curie remarked: "Imagine a young man of twenty-one living to be three times the normal human period of existence, retaining at the age of 210 years his youthful appearance at 21!"[71] Radium, the "pocket edition of the Roentgen tube,"[72] was long seen as having a mysterious and possibly beneficial effect on longevity.

Thus, in these earliest years, the hopes attached to the use of x rays encompassed both poles of the dialectic of medical futurism. On the one hand, physicians hoped to use them to extend and consolidate the gains of cooperative, preventive medicine by killing the germs and bacteria that spread contagious disease and by diagnosing those who carried such diseases. On the other, the rays represented the miracle cure that someday, with the flick of a switch, might heal a wide range of mortal ills. One author called the field of radiology a "veritable fairyland of science" in which even the most extravagant hopes might someday be realized.[73]

Moreover, some physicians viewed the advent of x rays as a signal that medicine's machine age had finally arrived. X rays were the product of a real, concrete, scientific machine; this was perhaps the source of their greatest promise. In contrast with the ill-defined and complex workings of cooperative preventive medicine, dreamers could look to a future in which machines would eliminate the work of public-health planners. Through technology, physicians would achieve the miraculous—and, in an age that believed in material progress more than in anything else, new and better machines seemed only a step away.

Physicians were not alone in their hopes for a miracle machine. In early 1897 one author noted that it was understandable that medical personnel would be enchanted with the new x-ray technology but that "the unbounded enthusiasm of the unscientific multitude was unlooked for, and [required] a special explanation."[74] His own explanation focused on the more sensational aspects of the x-ray apparatus—the flashing light and the shadowy pictures, he believed, were what fascinated the public.

But there was more. A deeper change in the image of medicine in the popular imagination can be seen in the popular fiction of the period, especially the utopian works. A sampling of these works from 1895 through 1906 shows a radical change in the types of medical innovations hoped for—a shift from the Victorian ideal of cooperative public health to a twentieth-century vision of machine-made health.

Almost immediately after their discovery, x rays began to appear in futuristic and utopian fiction. Although the vision of a reorganized society did not suddenly vanish in 1896, the machine (especially the x-ray machine) became an important and sometimes a dominant theme in literary speculations about the future of medicine. In a little-known story published in 1898 called "With the Eyes Shut," Edward Bellamy envisioned machines that would extend the senses, the strength, and the "second-sight" of their users.[75] During the next ten years, writers of futuristic fiction used the rays again and again to depict the possible extension of healing powers and supernatural capabilities. In addition, the principle of x rays as an active form of energy was later translated into a means for everything from space flight to warfare; the tradition that gave rise to "ray guns" and Superman's "x-ray vision" began in the first years of this century.

Like contemporary scientific investigators, dreamers in prose had high hopes that x rays would cure contagious diseases—specifically, that they would be used as a bacteriocide. In *Great Awakening, The Story of the Twenty-Second Century* (1899), A. A. Merrill's protagonist dies in 1901 and reawakens in a socialist utopia in 2199.[76] At one point in his travels he meets a biologist who, in working with "etheric" and other light waves, has discovered a ray fatal to the "tuberculosis germ." Merrill displays an interesting knowledge of the drawbacks of the rays in his description of tuberculosis treatment in his imaginary future: "The patient is subjected to the rays from this instrument for short periods of time until all the germs, in whatever part of the body, are killed; but these rays, being detrimental to the cellular tissue, have to be handled with great care, and the patient has to undergo another treatment to overcome these bad effects."[77] In a later work, W. S. Harris's *Life in a Thousand Worlds* (1905), the time traveler to the planet Dore-lyn finds that open-heart and lung surgery are performed under a "sheet of thermal rays" to prevent infection.[78]

Others saw the obvious possibilities of using the diagnostic x-ray technique to expand the range of human vision. In Jack Adams's *Nequa or the Problem of the Ages* (1900), a woman disguised as a man travels to a utopian region near the North Pole. In this land, Altruria, the visitor is told about an invention that provides vision through opaque matter, "an instrument that would cast the reflection on the retina of the eye."[79] This device has been applied to aerial surveys and other activities where sight through solid objects was required. Most important, though, these "electromagnetic optical instruments" allow physicians "to look into the bodies of their patients and examine the internal organs."[80]

Popular fiction also reflected contemporary interest in the use of x rays to send and see "thoughts." Merrill's *Great Awakening* shows a characteristic concern with the relationship between a sixth sense and rays of light and thoughts "vibrating in the ether."[81] The time traveler in Harris's *Life in a Thousand Worlds* visits Ploid, "the planet of highest invention," and discovers "thought photography."[82] Although thought flashes and telepathy occur repeatedly in the other worlds of Harris's novel, the technique is most highly developed on Ploid. There the inhabitants possess an "ability to follow the course of thought in a living cerebrum after the brain has been made visible by a light more potent than the X-ray." "After this exposure, the operator, with his wizard magnifying lens, watches the tiny tremulous brain cells in their infinitesimal quivering, as they carry messages from the soul to the world of sense and being."[83]

In the new utopias proposed at the turn of the century, technology, especially electrical invention, was central to the vision of the future. In William Alexander Taylor's futuristic land of *Intermere* (1901–02), the visitor is reminded that in 1900 the "current of the Universe" had been viewed with some awe and apprehension, but that to the technologists of the future it is "as gentle and harmless as the flowers that bloom by the wayside."[84] Intermere has no physicians and no medicine; those had been left behind: "Science and scientific discovery, as we utilize and employ them, have freed us from disease and made death but the exchange of lives."[85] Although this island paradise is illuminated by cathode-type lamps of various descriptions, "scientific" discovery has eliminated the problem for which they were used in Taylor's own time and has provided limitless life and health.

The elimination of death in the future was a favorite theme in the new scientific utopias, echoing much older utopian constructions. In John Ira Brant's *The New Regime: A.D. 2202* (1909), an amnesiac travels about, being reeducated into the ways of twenty-third-century America. In a land where the War Department has been replaced by such governmental branches as Music, the Stage, and Scientific Research, all things seem possible.[86] On a visit to the national biological laboratories in the city of Faraday, the amnesiac is told that biologists are working on a process by which a person's life can be held stationary at any desired point.[87] In Thomas Kirwan's *Reciprocity (Social and Economic) in the Thirtieth Century: The Coming Cooperative Age* (1909), a traveler leaves Boston in 1907, falls asleep, and arrives at his destination in 2907. Although much of his attention is given to social organization, machines and solar power play a great part in the story. Long life has been attained through applications of both science and the "sun's rays."[88]

Not all utopian novels of the 1890s and the 1900s concerned themselves with medical technology, but even many of those that focused on social reform included references to x-ray-related matters. Sometimes fictional travelers in these utopias encountered substances such as "etherine," invented by a physicist in Simon Newcomb's *His Wisdom*

The Defender (1900). This substance emits rays that act on "the ether of space" to allow etherine-coated objects to fly through the air and even through space.[89] Newcomb was a scientist, and his work can be contrasted with the thoroughly non-technology-oriented utopia of *The World a Department Store: A Story of Life Under a Cooperative System*, by Bradford Peck (1900). In this offbeat book, a proposed socialist society for 1925 is compared to a large urban department store.[90] The author was not concerned with medicine except to see it equitably distributed. H. G. Wells indulged in little but "preventive" speculation about future medicine in *A Modern Utopia* (1905).[91] Thus, not all these futurist authors wrote about the medical machine. Even the least technological social utopias, however, sometimes showed the influence of x rays and radium. In Cosimo Noto's *The Ideal City* (1903), the socialistic New Orleans of the future twinkles brightly at night with the reflection of radium streetlights on zinc sulphide houses.[92]

It is not possible to draw a sharp dividing line between the beginning and the end of an attitude about the future, or to explain all the various reasons for a shift in a popular viewpoint. In one case, however, it is possible to see such a shift represented in the work of a single author, and to look at the differences in perception evidenced by this shift. The impact of x rays as a popular symbol of medical and technological achievement can be seen clearly in the difference between two books written in 1892 and 1903 by John MacMillan Brown under the pseudonym Godfrey Sweven.

In the earlier book, *Riallero: The Archipelago of Exiles* (written in 1892 but not published until 1897), a traveler on a Swiftian journey observes various types of social and political organization. The tone is often caustic, and the book is a treatise on the limits, necessities, and abuses of centralized power in society.[93] Although well written, the volume has very little to distinguish it from any of the other socially conscious post-Bellamy utopias or dystopias.

In 1903, more than seven years after the discovery of x rays, Brown published another novel, which was advertised as a sequel. However, *Limanora: The Island of Progress* is a completely different kind of work. Limanora is a technological utopia as well as a paradise of the spirit. So complex is the technology presented that Brown provides the reader with a "vocabulary" of 88 common Limanoran terms and inventions. Among these are the following:

alclirolan—radiographic cinematograph; an instrument combining microscope, camera in vacuo, and electric power
airolan—a sensometer, or instrument for finding the personal equation of man

lavolan—revealer of the inner tissues and mechanisms
mirlan—life-lamp for revealing and recording internal processes for the use of the eye, the ear, and the electric sense[94]

Electricity, and especially the interaction of electricity and light, is the central motif of technology in this utopia. There are "electrographs" for taking exact readings of dreams, "historoscopes" for projecting moving pictures of history on the walls, and even a special profession for those who (by their "electric sense") imagine and project the distant future.[95] X-ray-derived devices are used for everything from cementing the bonds of true love to prolonging life. The alclirolan, a kind of celestial x-ray machine, is all powerful; "there was no new disease or microbe but gave up its secrets to this instrument."[96] Machines solve all medical problems. Between 1892 and 1903, Brown had radically changed his view of the future from one inspired by social and political theory to one predicated on scientific and technological change, with the x-ray machine as the primary model for the expansion of scientific capabilities. Brown had moved from a detached, often sarcastic stance as a social critic to an optimistic embrace of "science fiction." He had become an active dreamer about the future role of the machine—especially the medical machine.

As the twentieth century unfolded, imaginative fiction continued to reflect the shift in medical futurism brought about by the invention of the x-ray machine. In science fiction, machines have customarily expanded medical vision and extended the duration and the quality of life. Yet the x-ray machine's influence was much more general. As we have seen, it provided new models for thinking about medical treatment and new reasons for hope. Once called a "light in dark places"[97] because of its ability to probe the unseen interior of the human body, the x-ray machine led the way to the technical, machine-oriented medical practice we are familiar with today. Although its primacy as a therapeutic device has been challenged by even more "miraculous" machines, the hopeful beam it first cast nearly a century ago toward a future of total health still illuminates our lives.

Notes

1. For a discussion of the earliest views of perfect health, see G. Kasten Tallmadge, "The Physician in the Ancient Utopias," *Ciba Symposia* 7 (1945), pp. 166–177.

2. Benjamin Ward Richardson, *Hygeia: A City of Health.* A Presidential Address Delivered Before the Health Department of the Social Sciences Association at the Brighton Meeting, October 1875 (London: Macmillan, 1876), p. 10.

3. Thomas More's analysis of utopian medical care is discussed in G. Kasten Tallmadge's "Medicine in the Utopias of the Renaissance and the Seventeenth Century," *Ciba Symposia* 7 (1945), pp. 178–187.

4. Richardson, *Hygeia*, p. 10.

5. Edward Bellamy, *Looking Backward, 2000–1887*, originally published in 1888 (New York: New American Library, 1960), p. 137.

6. Ibid., p. 173.

7. Chauncey Thomas, *The Crystal Button: Or, Adventures of Paul Prognosis in the Forty-Ninth Century*, originally published in 1891 (Boston: Gregg, 1978), p. 67.

8. A prominent physician, Zabdiel Boylston Adams, recalls this viewpoint in "An Epoch in Medicine in an Age of Delusion," *Boston Medical and Surgical Journal* 136 (1897), p. 585.

9. N. Pearson, *National Progress and National Character* (New York: author, 1894), p. 67.

10. H. Elsdale, "Scientific Problems of the Future," *Annual Report of the Board of Regents of the Smithsonian Institution, 1894* (Washington, D.C.: U.S. Government Printing Office, 1896), p. 679.

11. J. P. Barrett, *Electricity at the Columbian Exposition* (Chicago: Donnelley and Sons, 1894), p. 435.

12. The x ray was called a variety of names during its first years of popularity. To electricians it was the "new light," to photographers the "new photography," and to physicians the "new medicine."

13. The best English translation is the facsimile edition in Herbert Klickstein's folio *Wilhelm Conrad Röntgen: 'On a New Kind of Ray': A Bibliographical Study* (New York: Mallinckrodt Chemical, 1966).

14. The most straightforward accounts of the discovery may be found in George Sarton's "The Discovery of X-rays, *Isis* 26 (1937), pp. 340–369; Alan Ralph Bleich's *The Story of the X-rays from Roentgen to Isotopes* (New York: Dover, 1961); and Otto Glasser's *Doctor W. C. Röntgen* (Springfield, Ill.: Charles C. Thomas, 1965).

15. "Roentgen or X-ray Photography," *Scientific American* 74 (February 15, 1896), p. 103.

16. "The New World of Science," *Scientific American* 74 (February 15, 1896), p. 98.

17. "Tests with New Plates . . . ," *New York Times*, February 14, 1896, p. 9, col. 4.

18. "Hidden Solids Revealed. Prof. Routgen Experiments with Crooke's Vacuum Tube," *New York Times*, January 16, 1896, p. 9, col. 5.

19. "The Rontgen Discovery," *New York Times*, January 29, 1896, p. 9, col. 1.

20. "Some Startling Results at Yale . . . ," *New York Times*, February 5, 1896, p. 9, col. 2.

21. This double cartoon by Charles Dana appeared in *Life* 27 (April 1896), p. 313.

22. "X-ray Advertising," *American Journal of Photography*, August 1896, pp. 383–385.

23. *Electrical Review*, February 17, 1897, p. 73.

24. Glasser, *Doctor W. C. Röntgen*, p. 203.

25. See "Commercial Application of X-rays," *Electrical World* 27 (March 1896), p. 339.

26. "Photography Up to Date," *Electrical Engineer* 22 (September 1896), p. 253.

27. "Some Remarkable Electrical Applications," *Electrical World* 28 (September 26, 1896), p. 365.

28. Editorial, *Electrical Engineer* (January 1896), p. 258.

29. Bleich, *Story of the X-rays*, p. 32.

30. "Funny Ideas About the Roentgen Rays," *Electrical Engineer* 21 (April 1896), p. 377.

31. "X-ray Slot Machines," *Wilson's Photography Magazine* 39 (March 1902), p. 54.

32. See editorial, *Photographic Times* 28 (November 1896), p. 493.

33. Cartoon, "Another Application of Roentgen's Discovery—Baseball Player's Glass Arm," *Western Electrician* 18 (February 1896), p. 87.

34. "A Preliminary Report on the Roentgen Ray," *Journal of the American Medical Association* 26 (February 1896), pp. 402–404; "The Roentgen Ray as a Moral Agent," ibid. 26 (April 1896), p. 843.

35. "To a Fickle Miss," *Life* 27 (February 1896), p. 151.

36. "The Roentgen Ray and Second Sight," *Journal of the American Medical Association* 26 (May 1896), p. 1065.

37. David Starr Jordan, "The Sympsychograph: A Study in Impressionist Physics," *Appleton's Popular Science Monthly* 49 (September 1896), pp. 597–602.

38. Ibid., p. 597.

39. "Sympsychograph," *Electrical World* 28 (Ocotober 1896), p. 403.

40. "The Philosopher's Stone at Last," *Electrical Engineer* 21 (May 1896), p. 472.

41. "Edison's New Electric Light," *Scientific American* 74 (June 1896), p. 378.

42. Phillip Mills Jones, "X-rays and X-ray Diagnosis," *Journal of the American Medical Association* 29 (November 1897), p. 946.

43. "Hidden Solids Revealed," *New York Times*, January 16, 1896, p. 9, col. 5.

44. "Demonstration of the Roentgen Rays," *Boston Medical and Surgical Journal* 134 (April 1896), pp. 449–450.

45. Edward P. Davis, "The Study of the Infant's Body and of the Pregnant Womb by the Roentgen Ray," *American Journal of the Medical Sciences*, n.s., 111 (March 1896), pp. 269–270.

46. See "The Roentgen Rays in Surgery," *Journal of the American Medical Association* 26 (March 1896), p. 548; "An Application of the Rontgen Rays to Intrauterine Photography," *American Journal of the Medical Sciences*, n.s., 111 (June 1896), pp. 739–740.

47. "Radiography," *Journal of the American Medical Association* 30 (February 1898), p. 337.

48. "Thomas A. Edison's Experiments, " *New York Times*, February 8, 1896, p. 9, col. 7; "Direction of the X-rays," ibid., February 11, 1896, p. 16, col. 1.

49. "The Employment of X-rays for the Relief of Some Forms of Blindness," *Journal of the American Medical Association* 28 (January 1897), p. 180; "The Roentgen Rays and the Blind," ibid. 30 (May 1898), p. 1307.

50. Louis Bell, "Effect of Roentgen Rays on the Blind," *Electrical World and Engineer* 28 (December 1896), p. 729.

51. "Direction of the X-Ray," *New York Times*, February 11, 1896, p. 16, col. 1.

52. "Bacilli and X-Rays . . . ," *New York Times*, March 15, 1896, p. 9, col. 7.

53. "The Roentgen Rays in Tuberculosis," *Journal of the American Medical Association* 31 (December 1898), p. 1437.

54. See, for instance, one of the first reports of lupus treatments with the rays: "Case of Lupus," *Journal of the American Medical Association* 33 (July 1899), pp. 162–163.

55. J. William White, "Surgical Application of the Roentgen Rays," *American Journal of the Medical Sciences*, n.s., 115 (January 1898), p. 14. See also "Tuberculosis Peritonitis Apparently Cured by the X-ray," ibid. 117 (February 1899), p. 222.

56. "Bacteriocidal Action of Roentgen Rays," *American Journal of the Medical Sciences*, n.s., 124 (July 1902), p. 163.

57. "Effect of X-rays on the Skin," *Electrical World and Engineer* 28 (December 1896), p. 761.

58. "Action of the Roentgen Rays Upon the Normal Skin and Hair Follicles," *American Journal of the Medical Sciences*, n.s., 113 (May 1897), p. 590.

59. "Roentgen Rays in the Treatment of Skin Diseases and for the Removal of Hair," *American Journal of the Medical Sciences*, n.s., 121 (January 1901), p. 121.

60. "X-Ray Burns," *Journal of the American Medical Association* 38 (January 1902), p. 278.

61. See, for instance, G. Betton Massey, "The Increasing Prevalence of Cancer as Shown in the Mortality Statistics of American Cities," *American Journal of the Medical Sciences*, n.s., 119 (February 1900), pp. 170–177.

62. White, "Surgical Applications."

63. Emil Grubbe (1875–1960) claimed to have been the first to use x rays in the treatment of cancer. His claim was widely disputed in the medical field.

64. For a representative article on the use of x rays to treat cancer, see J. F. Rinehart, "The Use of the Rontgen Rays in Skin Cancer, Etc., With Report of a Case," *American Journal of the Medical Sciences*, n.s., 124 (July 1902), pp. 115–119.

65. M. M. Guilloz and Spilliman, "The Roentgen Rays in Splenic Leukemia," *American Journal of the Medical Sciences*, n.s., 128 (September 1904), p. 543.

66. James P. Marsh, "A Case of Supposed Sarcoma of the Chest Wall Symptomatically Cured by Means of the X-ray," *American Journal of the Medical Sciences*, n.s., 127 (June 1904), pp. 1054–1056.

67. Charles W. Allen, "The X-Ray in Cancer and Skin Diseases," *Journal of the American Medical Association* 40 (February 1903), p. 509.

68. Arthur Dean Bevan, "The X-Ray as a Therapeutic Agent, with Especial Reference to Carcinoma," *Journal of the American Medical Association* 42 (January 1904), p. 29.

69. "Nikola Tesla Discusses X-Rays . . . ," *New York Times*, March 11, 1896, p. 16, col. 1; "Views on Tesla's Ideas," ibid., March 12, 1896, p. 9, col. 3; "Roentgen Rays and Material Particles," *Electrical World and Engineer* 28 (August 1896), p. 127.

70. "Tests With New Plates," *New York Times*, February 14, 1896, p. 2, col. 4.

71. "Radium and Longevity," *Journal of the American Medical Association* 43 (August 1904), p. 617.

72. "Foreign News," *Journal of the American Medical Association* 44 (January 1905), p. 229.

73. Elihu Thomson, "Electricity in the Coming Century," *Electrical World and Engineer* 37 (January 1901), p. 21.

74. "A Review of the Year 1896," *Electrical World and Engineer* 29 (January 1897), p. 3.

75. Edward Bellamy, *The Blindman's World and Other Stories*, orig. pub. 1898 (New York: Garrett, 1968).

76. Albert Adams Merrill, *The Great Awakening: The Story of the Twenty-Second Century* (Boston: George, 1899).

77. Ibid., p. 294.

78. W. S. Harris, *Life in a Thousand Worlds*, orig. pub. 1905 (New York: Arno, 1971), p. 212.

79. Alcanoan Q. Grigsby (pseud. Jack Adams), *Nequa or the Problem of the Ages* (Topeka: Equity, 1900), p. 141.

80. Ibid., p. 141.

81. Merrill, *Great Awakening*, p. 119.

82. Harris, *Life in a Thousand Worlds*, p. 236.

83. Ibid., p. 236.

84. William Alexander Taylor, *Intermere*, orig. pub. 1901–02 (New York: Arno, 1971), p. 69.

85. Ibid., pp. 117–118.

86. John Ira Brant, *The New Regime: A.D. 2202* (New York: Cochrane, 1909), p. 35.

87. Ibid., p. 64.

88. Thomas Kirwan (pseud. William Wonder), *Reciprocity (Social and Economic) in the Thirtieth Century: The Coming Cooperative Age* (New York: Cochrane, 1909), p. 40 and passim.

89. Simon Newcomb, *His Wisdom The Defender. A Story.* (New York: Harper and Bros., 1900), p. 21.

90. Bradford Peck, *The World a Department Store: A Story of Life Under a Cooperative System* (Boston: author, 1900), p. 70.

91. H. G. Wells, *A Modern Utopia* (New York: Charles Scribner's Sons, 1905), pp. 314–315.

92. Cosimo Noto, *The Ideal City* (New York: author, 1903), p. 242.

93. John MacMillan Brown (pseud. Godfrey Sweven), *Riallero: The Archipelago of Exiles* (New York: G. P. Putnam's Sons, 1897).

94. John MacMillan Brown (pseud. Godfrey Sweven), *Limanora: The Island of Progress* (New York: G. P. Putnam's Sons, 1903), pp. vii–ix.

95. Ibid., p. ix.

96. Ibid., p. 130.

97. "A Light in Dark Places, Roentgen's Photographs Through Flesh and Through Wood," *New York Times*, January 30, 1896, p. 9, col. 5.

2

Amateur Operators and American Broadcasting: Shaping the Future of Radio

Susan J. Douglas

"If any of the planets be populated with beings like ourselves," wrote William Preece in 1898, "then if they could oscillate immense stores of electrical energy to and fro in telegraphic order, it would be possible for us to hold commune by telephone with the people of Mars."[1] Preece, the engineer-in-chief of the British Post Office and a radio experimenter, was one of the first of many to predict that intergalactic "wireless" signaling would, in the near future, connect us with others across the great expanse of the universe. His comments, circulated in American magazines and newspapers, were reiterated by some and ridiculed by others. But although Preece's heady vision may have seemed farfetched to some, it was only the ultimate extension of other more earthbound yet highly enthusiastic predictions surrounding radio at the turn of the century. From its first public unveiling and through the next 25 years, the invention evoked a range of prophecies—some realistic, some fantastic, and nearly all idealistic—of a world improved through radio.

Radio both fitted into and extended Americans' notions of how the future would be made better, maybe even perfect, through technology. Not all technology was so embraced. The factory system, with its large, noisy, seemingly autonomous machines, had produced a range of complicated social problems that profoundly frustrated most Americans. At the turn of the century, the American press was filled with self-congratulatory assessments of how far the country had come in 100 years, but just beneath that veneer of optimism was a deep anxiety about the dislocations and vulnerabilities that had accompanied industrialization. Radio was not only exempt from this anxiety, it was also meant to relieve it. Like certain other inventions, radio was seen as delivering society from a troubled present to a utopian future. What was it about radio that evoked both idealistic and fantastic visions of

Figure 1
Frontispiece of Hugo Gernsback's *Radio for All* (1922), showing what radio might be expected to do in the future.

the future? What influence, if any, did these highly publicized and richly embellished predictions exert? Who, if anyone, believed them? Did they influence the course of broadcasting's early history?

Although other scientists and inventors had been experimenting with the wireless transmission of electrical energy and signals, it was Guglielmo Marconi, an Irish-Italian inventor, who first sought to make such signaling practicable and commercially available. In 1896 he introduced British government officials to his method of sending dots and dashes through "the air." Marconi intended to provide point-to-point communications between ships and between ship and shore. He called his invention a wireless telegraph, and he saw it as an adjunct to and a potential competitor with the telegraph and the transatlantic cables. Thus, radio began not as a method of broadcasting speech and music to a vast audience but rather as an analogy to and an extension of the telegraph. Telegraphy had revolutionized first domestic and then international communications, but its reliance on a network of wires physically limited the range and scope of its use. In remote areas, on islands, or aboard ships, where wires were not or could not be connected, people were unable to take advantage of the new invention. Thus, there were still potential customers left out of the communications revolution.

In 1899 Marconi brought his apparatus to the United States and demonstrated its potential for ship-to-shore signaling and for news-gathering during the highly publicized America's Cup races. Having installed his invention aboard a small ship, he followed the progress of the race and "wirelessed" the day's results to newspaper reporters on shore. News of his achievements inspired awe and wonder; to people still getting accustomed to the telegraph, cables, and the telephone, signaling without any wires—without any tangible connection at all—was fantastic and incredible.

Wireless telegraphy made its debut at the close of what the press often referred to as the "Century of Science." Invention was celebrated in the nineteenth century as the handmaiden of cultural progress; inventors were exalted as the discoverers and interpreters of nature's secrets. Popular magazines showed little restraint in describing recent advances in scientific and technical knowledge. Technological progress and social progress seemed intertwined in an ineluctable upward spiral. In the words of a *Scientific American* writer, "The railroad, the telegraph, and the steam vessel annihilated distance; people touched elbows across the seas; and the contagion of thought stimulated the ferment of civilization until the whole world broke out into an epidemic of industrial progress."[2] While words such as *contagion* and *epidemic* suggested that industrialization was like a disease, the implied cure was better inventions more in tune with natural forces. Noting that the scientific movement was "at its maximum of vigor and productiveness," *Popular Science Monthly* nonetheless felt that, because of the recent advances in scientific inquiry, nature would "go on revealing herself to us with greater and greater fullness."[3] These were expressions of an America that still believed the right technical advances would bring a utopia closer. Those advances that unraveled nature's mysteries and tied them to practical applications found particularly warm reception in the popular press.

Reporters responded to wireless telegraphy with unprecedented awe. On December 15, 1901, when Marconi reported to the press that he had successfully transmitted the letter S from England to Newfoundland, he garnered bold front-page headlines and effusive praise. The press lionized the inventor-hero and compared him to Edison.[4] Popular magazines sent reporters to interview him and featured illustrated stories detailing in often melodramatic style the delays, doubts, and hardships that had preceded his success. With optimistic and excited rhetoric, these articles celebrated the new invention. "Our whole human existence is being transformed by electricity," observed the *North American Review*. "All must hope that every success will attend

Marconi and the other daring adventurers who are exploring this comparatively unknown scientific region."[5] Success was so important, continued the magazine, because no invention was "more pregnant with beneficial possibilities or calculated to be a more helpful factor in advancing the existing order of the world's life." *Current Literature* declared: "Probably no other modern scientific discovery has had so much romantic coloring about it as wireless telegraphy."[6] Wireless held a special place in the American imagination precisely because it married idealism and adventure with science: "The essential idea belongs to the realms of romance, and from the day when the world heard with wonder, approaching almost incredulity, that a message had been flashed across the Atlantic . . . the wonder of the discovery has never decreased."[7] Ray Stannard Baker, writing for *McClure's*, tried to transport his readers to this romantic realm by putting them in Marconi's position: "Think for a moment of sitting here on the edge of North America and listening to communication sent *through space* across nearly 2000 miles of ocean from the edge of Europe! A cable, marvelous as it is, maintains a tangible and material connection between speaker and hearer; one can grasp its meaning. But here is nothing but space, a pole with a pendant wire on one side of a broad, curving ocean, an uncertain kite struggling in the air on the other—and thought passing between."[8] *World's Work* asserted: "The triumph of Marconi remains one of the most remarkable and fruitful that have ever crowned the insight, patience, and courage of mankind."[9]

Celebration quickly led to prediction. In contrast with the forecasts for other technologies, however, there were few, if any, forecasts of how wireless equipment would look or how it would change the American landscape. There were no speculations on wireless sets of the future, no fantastic drawings of modernistic equipment. Rather, the predictions focused on where the messages might go and on what wireless would do for society and for individuals.

Reporters, often repeating the pronouncements of inventors eager to sell apparatus, began anticipating how wireless would make the world a happier, less risk-filled, and more civilized place. Some prophecies, though overenthusiastic, were realistic and quickly came true—for instance, the following: "On shipboard a word from shore or from another ship will give warning of a neighboring reef or an ice floe, or tell of a dangerous change in an ocean current or herald a coming tempest. Steamer may communicate with steamer throughout the whole course of an ocean lane, forfending all risk of collision."[10] P. T. McGrath, a newspaper editor who interviewed Marconi shortly after his transatlantic experiment, predicted in *Century Magazine* that wireless would "play a great part" in future military operations and would facilitate the exploration of places as remote and forbidding as the "arctic solitudes."[11] These forecasts reflected the early marketing goals of those who, like Marconi, sought to model wireless after the telegraph system and thus offer a similiar but competing communications network. But such a circumscribed vision, focusing on the existing market, could not contain the hopes and dreams wireless evoked. The possibilities seemed so much grander, and they extended beyond the advantages that might be enjoyed by ships' passengers, soldiers, and explorers. In the eyes of the press, wireless held the promise of being a truly democratic invention, benefiting millions instead of a few hundred.

The introduction of wireless renewed hopes for the possibility of eventually securing world peace. *Popular Science Monthly* observed: "The nerves of the whole world are, so to speak, being bound together, so that a touch in one country is transmitted instantly to a far-distant one."[12] Implicit in this organic metaphor was the belief that a world so physically connected would become a spiritual whole with common interests and goals. The *New York Times* added: "Nothing so fosters and promotes a mutual understanding and a community of sentiment and interests as cheap, speedy and convenient communication."[13] Articles suggested that this technology could make men more rational; with better communications available, misunderstandings could be avoided. These visions suggested that machines, by themselves, could change history; the right invention could help people overcome human foibles and weaknesses, particularly rivalry and suspicion brought on by isolation and lack of information.

The most stirring prophecies, however, envisioned individual rather than social benefits. A sonnet published in the *Atlantic Monthly*, entitled simply "Wireless Telegraphy," traced the flight of a word "over the wilds of ocean and of shore" until it reached its intended destination:

Somewhere beyond the league-long silences,
Somewhere across the spaces of the years,
A heart will thrill to thee, a voice will bless,
Love will awake and life be perfected![14]

Love and life would be "perfected" as wireless communication would ease loneliness and isolation. The *New York Times* foresaw a time when "wireless telegraphy would make a father on the old New England farm and his son in Seattle . . . neighbors—perhaps by the use of their own private apparatus."[15] *The Century Magazine*, reporting "vastly greater things are predicted for the future" of wireless, offered this rather poignant prophecy:

> . . . if a person wanted to call to a friend he knew not where, he would call in a very loud electromagnetic voice, heard by him who had the electromagnetic ear, silent to him who had it not. "Where are you?" he would say. A small reply would come "I am at the bottom of a coal mine, or crossing the Andes, or in the middle of the Atlantic." Or, perhaps in spite of all the calling, no reply would come, and the person would then know that his friend was dead. Think of what this would mean, of the calling which goes on every day from room to room of a house, and then think of that calling extending from pole to pole, not a noisy babble, but a call audible to him who wants to hear, and absolutely silent to all others. It would be almost like dreamland and ghostland, not the ghostland cultivated by a heated imagination, but a real communication from a distance based on true physical laws.[16]

The rhetoric of this vision gets at the heart of what excited people about wireless. It was the potential autonomy and spontaneity of such communication that gripped Americans' imaginations: People could talk to whomever they wanted whenever they wanted, no matter how much distance or how many obstacles intervened. This technology would help them transcend the social and economic forces—particularly, and ironically, industrialization—that had driven them apart.

America's population was in the midst of an accelerated shift from country to city living, resulting in more frequent separation of family members and friends. To communicate over distances, individuals could either write letters or go through corporate intermediaries. Using the networks operated by the Bell Company or Western Union could be expensive, and privacy was compromised; both the telegraph and the telephone relied on operators, who were, either by necessity or inclination, privy to the contents of the message. Wireless seemed to promise something different: instant communication, through "the air," free from both operators and fees. In addition, wireless seemed the technical equivalent of mental telepathy. Intelligence could pass between sender and receiver without tangible connection. Thus, to many, wireless bridged the chasm between science and metaphysics, between the known and the unknown, between actual achievement and limitless possibility.

Though there were few predictions about how the wireless apparatus of the future would look, one recurring image was that of the "pocket-sized" wireless kit that could be taken and used anywhere.[17] The small size of such a device and its detachment from a corporate network contributed considerably to its appeal. Finally, the articles suggested, here was an invention people themselves could handle and control. Thus it was very different indeed from other technologies, such as factory machinery, railroads, the telegraph, or the telephone. If a better future was to be attained through particular technologies, and if this invention was one that all could master, then couldn't individuals, through radio, shape their own futures? This was one of the unarticulated yet central aspects of the allure of wireless.

Why did the press print and encourage all these prophecies, naive and farfetched as some of them seemed? The success of utopian novels in the late nineteenth century indicated that visions of the future, which offered escape from the problems of the present, were what the public wanted. Such popularity was one of the keys to the continued appearance of prophecy in journalism. Circulation wars among newspaper publishers during the 1880s and the 1890s contributed to an increased reliance on sensationalism in headlines, stories, and illustrations to attract yet more readers. Given the prevailing celebration of technological progress, which was perpetuated by the press, reporters looked for scoops on the latest technical feat or for predictions of feats to come. The grander the prophecy, the more dramatic (and beneficial) the changes envisioned, the better; caution did not sell newspapers.[18]

Most important, newspaper publishers sought to promote wireless because they believed the invention would accelerate and cheapen newsgathering. The press considered the transatlantic cable rates for foreign news onerous; the *New York Times* once referred to the cable companies as "monopolistic serpents."[19] Marconi predicted that wireless would shortly provide a transatlantic news service at much cheaper rates than those imposed by the cable companies, and the press reiterated this forecast frequently.[20] One reporter said simply, "Cables might now be coiled up and sold for junk."[21] Thus, there were economic as well as social and cultural stimuli generating the proliferation of these visions.

However, the enthusiasm of the press was not matched in either the corporate world or the armed services, for there was still a great discrepancy between present potential and predicted promise in the wireless field. The appealing predictions in the newspapers and magazines envisioned invisible, point-to-point connections between an unlimited number of senders and receivers. Implicit in these forecasts was the belief that there would be room for all potential users in "the air" and that these point-to-point conversations would be private. But in reality, Marconi and others were still struggling to eliminate interference between messages and ensure secrecy. These disadvantages, which inventors kept predicting would soon be corrected, explain the hesitancy of military and commercial clients to purchase wireless equipment. In addition, a combination of erratic equipment performance, poor marketing strategies, and corporate indifference or wariness about the potential value of wireless in general left the invention with no definite niche in the American marketplace. The two major communications companies, Western Union and American Telephone and Telegraph, concentrated on preserving the hegemony of their own systems and did not see any immediate advantage to acquiring and promoting the new technology. Only the most prestigious ocean liners, such as those of the Cunard line, installed wireless; smaller steamship companies were slow to adopt the invention. The American inventors Lee DeForest and Reginald Fessenden, who made the most significant technical contributions to American radio development, failed to find steady customers and a regular clientele. From a business standpoint, wireless was a failure.

Yet from 1906 to 1912, as American wireless companies were on the verge of or in fact declared bankruptcy, the United States experienced its first radio boom. Thousands of people, believing in a profitable future for the invention, bought hundreds of thousands of dollars' worth of stock in fledgling wireless companies.[22] Others took even more decisive action and began to construct and use their own wireless stations. Thus, while the leaders of American corporate and bureaucratic institutions regarded the various prophecies with a skeptical eye, other individuals began translating vision into action. These Americans—primarily white middle-class boys and men who built their own stations in their bedrooms, attics, or garages—came to be known as the amateur operators, and by 1910 their use of wireless was being described in newspapers and magazines around the country. *The Outlook* outlined the emerging communications network:

> In the past two years another wireless system has been gradually developing, a system that has far outstripped all others in size and popularity. . . . Hundreds of schoolboys in every part of the country have taken to this most popular scientific fad, and, by copying the instruments used at the regular stations and constructing apparatus out of all kinds of electrical junk, have built wireless equipments that in some cases approach the naval stations in efficiency.[23]

The amateurs were captivated by the idea of harnessing electrical technology to communicate with others, and were not deterred by lack of secrecy or interference. In fact, these features, considered such a disadvantage by institutional customers, increased the individual amateur's pool of potential contacts and the variety of information he could send and receive.

It is impossible to establish a causal relationship between the eager early predictions and these subsequent activities on the part of the amateurs. We cannot know how many wireless enthusiasts were inspired by what they read in newspapers or magazines. But we do know that there was a climate of enthusiasm, which the press reflected, embellished, and fanned. And we see the emergence of a widely dispersed group of individuals trying to accomplish what the journalistic visions had promised: communication over sometimes great distances with whomever else had a wireless set. The ways in which the amateurs came to use wireless—to contact strangers, to make friends, to provide communication during disasters, and to circumvent or antagonize private and governmental organizations—were enactments of the previously articulated visions.

The favorable social climate, though conducive to the development of the amateur wireless network, cannot fully explain the invention's proliferation. How were the amateurs able to master this particular technology? The first and most tangible development was the availability, starting in 1906, of the simple, inexpensive crystal set, a device that could, for some unexplained reasons, detect radio waves. Inventors did not understand how the crystal worked, but they knew that it was a sensitive, durable, inexpensive receiver that was simple to operate and required no replacement parts.[24] At the time, how and why the crystal worked was not as important as its simplicity and its very low cost. (The crystal was placed between two copper contact points, which were adjustable so that the pressure could be regulated and the most sensitive portion of the mineral selected. To keep the contact as small as possible, often a thin wire, known popularly as the "catwhisker," was used.[25]) The importance of the introduction of the crystal detector cannot be overemphasized. More than any other component, it contributed to the democratization of wireless, the concomitant wireless boom, and the radio boom of the 1920s.

The amateurs' ingenuity in converting a motley assortment of parts into working radio sets was impressive. With performance analogous to that of an expensive detector now available to them in the inexpensive crystal, the amateurs were prepared to improvise the rest of the wireless set. Before 1908 they had to, for very few companies sold equipment appropriate for home use. Also, one of the crucial components, the tuning coil, was not supposed to be available for sale because it was part of the patented Marconi system. As the boom continued, however, children's books, wireless manuals, magazines, and even the Boy Scout Manual offered diagrams and advice on radio construction. As one author instructed, "You see how many things I've used that you can find about the house."[26]

In the hands of the amateurs, all sorts of technical recycling and adaptive reuse took place. Discarded photographic plates were wrapped with foil and became condensers. The brass spheres from an old bedstead were transformed into a spark gap, and were connected to an ordinary automobile ignition coil cum transmitter. (Model T coils were favorites.[27]) Tuning coils were made out of curtain rods, baseball bats, or Quaker Oats containers wound with wire. One amateur described how he made his own rotary spark gap from an electric fan.[28] Another recalled that he "improvised a loudspeaker by rolling a newspaper in the form of a tapered cone."[29] Another inventor's apparatus was "constructed ingeniously out of old cans, umbrella ribs, discarded bottles, and various other articles."[30] Amateurs used umbrella ribs as well as copper or silicon bronze wire to erect inexpensive and fairly good aerials. The one component that was too complicated for most amateurs to duplicate, and too expensive to buy, was the headphone set. Telephones began vanishing from public booths across the United States as the amateurs lifted the earpieces for their own stations.[31] Thus, the amateurs didn't just adopt this new technology; they built it, experimented with it, modified it, and sought to extend its range and performance. They made radio their own medium of expression.

The size of this burgeoning wireless network is hard to gauge. Estimates vary, but Clinton De Soto, in his history of amateur radio, asserts that "it was the amateur who dominated the air."[32] In 1911, *Electrical World* reported: "The number of wireless plants erected purely for amusement and without even the intention of serious experimenting is very large. One can scarcely go through a village without seeing evidence of this kind of activity, and around any of our large cities meddlesome antennae can be counted by the score."[33] The *New York Times* estimated in 1912 that the United States had several hundred thousand active amateur operators.[34]

Although the availability of the crystal set was a prerequisite for the proliferation of wireless, the technology alone cannot account for the dramatic increase in the invention's use. Complementary cultural and social forces were at work, and their power becomes more apparent when we look at a very different sort of historical artifact: the contemporary popular culture. It is in the popular culture that the spirit and the zeal of amateur radio are revealed.

Articles and books published between 1907 and 1922 captured the enthusiasm and extended the visions first articulated in the press at the turn of the century. Radio was portrayed as an invention that provided entrée into an invisible realm unfamiliar to the less technically adventurous. This hint of exclusivity further romanticized the amateurs' activities while implying that they were helping to shape and bring about the future. In a feature article on the amateurs in November 1907, the *New York Times* evoked the mystery surrounding radio and hinted at what lay ahead: "For intrigue, plot and counterplot, in business or in love or science, take to the air and tread its paths, sounding your way for the footfall of your friends' or enemy's message. There is a romance, a comedy, and a tragedy yet to be written."[35] A more ethereal note was struck by Francis Collins in his children's book *Wireless Man*: "On every night after dinner . . . the entire country becomes a vast whispering gallery."[36] Most effusive of all was the assessment of radio offered by a character in the *Radio Boys* book series: "But honestly, I think radio is the greatest thing in this whole universe . . . What hasn't it done? What can't it do? . . . It's enough to make a dumb man eloquent."[37] It is difficult to establish how widely distributed these books glorifying radio's potential were, but a review of the *New York Times* index and the *Reader's Guide* indicates an escalating number of practical, popular, and visionary articles about the invention during this period.[38]

These breathless accounts of the adventures awaiting the enterprising enthusiast had special appeal for the young. Unlike their suspicious elders, who thought wireless too fantastic, too impractical, or too unremunerative, certain boys believed earnestly in the new marvel and were eager to explore its possibilities. Many were at the age when they could most easily learn a new technical "language." To them, radio promised excitement, fraternity, and new scientific knowledge. Through dime novels, vaudeville shows, and children's books, these youngsters witnessed, unhindered by any acquired disbelief, the unrefined and unself-conscious aspirations of the culture, especially the hope that technology would solve society's problems. Businessmen and military men were not part of this world, and they no doubt would have considered the fictional celebrations of wireless unrealistic.

By 1910, the prevailing theme in vaudeville and popular literature was man's mastery of and alliance with technology. This had supplanted western themes glorifying the white man's victory over the wild frontier. For an era that no longer had a frontier, the western motif became less compelling than the more urban, technological themes. A famous and popular vaudeville routine performed during the first decade of the century consisted of a team of men taking apart and reassembling a Ford on stage in 8 minutes. Youthful heroes of popular fiction, such as Frank Merriwell, Nick Carter, and Tom Swift, were able to meet any technologial challenge. Technology provided a new realm of adventure and conquest.

Fictional characters' lives were transformed with the advent of radio. *The Radio Boys with the Iceberg Patrol* described the exciting achievements of Bob and Joe, two amateur operators. With their radio sets, they track down a "rascal who had defrauded an orphan girl." They are "instrumental in rescuing people who had been run down by a stolen motorboat." Shortly after this, Bob and Joe "overhear and expose a scoundrelly plot of financial sharpers" and "secure the return to jail of desperate escaped convicts." Because their apparatus is small and portable, it goes everywhere with them. While relaxing at the shore, they learn of a "terrible storm" out at sea, and they are "able by a message to save the vessel on which their own people were voyaging."[39] All these adventures are recounted in just one of the book's action-filled paragraphs. In *Tom Swift and His Wireless Message*, Tom saves himself and his companions from a volcanic island by devising a wireless set and sending for help. "Would help come? If so, from where? And if so, would it be in time? These are the questions that the castaways asked themselves. As for Tom, he sat at the key clicking away, while, overhead, from the wires fastened to the dead tree, flashed out the messages." Finally, "from somewhere in the great void," a reply comes back and all are rescued.[40]

Early accounts, both fictional and journalistic, portrayed young radio amateurs as having the sort of physical features and personality traits previously extolled in the highly popular Horatio Alger stories. Bob, one of the Radio Boys, "was a general favorite because of his frank and sunny nature and his straightforward character. The elder people liked him, and among the younger element he was the natural leader. . . . He was tall for his age, of dark complexion and with eyes that always looked straight at one without fear or favor."[41] The *New York Times* described J. Willenborg, an amateur from Hoboken, New Jersey, as "grey-eyed, clear-cut of feature, intent, and with his brow furrowed." The reporter noted that although Willenborg's father was "well-to-do," Willenborg did not rely on his father's wealth. "He is so frequently called as an expert witness in so many important suits over electrical matters that his fees give him ample resources."[42]

The press emphasized that many of these industrious amateurs, after graduating from high school, became operators on ships and saved their salaries to pay for college or technical school. Contemporary celebrations of the "ambition and really great inventive genius of American boys"[43] no doubt lured new recruits. Parents were advised to encourage the hobby: "This new art does much toward keeping the boy at home, where other diversions usually, sooner or later, lead him to questionable resorts; and for this reason well-informed parents are only too willing to allow their sons to become interested in wireless.[44] As these various expressions of encouragement for the amateurs appeared, the amateurs, according to their reminiscences, took them to heart. As one amateur recalled, "We were undoubtedly romantic about ourselves, possessors of strange new secrets that enabled us to send and receive messages without wires."[45] This romantic self-image was heightened whenever real-life operators became heroes. In January 1909, when all but five passengers of the ships *Republic* and *Florida* were saved after the two ships collided in a thick fog, Jack Binns, the wireless operator, who had sent the distress signals, became a hero. Later, Binns was asked by vaudeville agents and book publishers to demonstrate his talents and tell his story.[46] He agreed to write the forewords for the Radio Boys books, and through these he emphasized that adventure and celebrity could happen to any boy at any time. While he acknowledged that "the escapades of the boys in this book are extremely thrilling," he added "but not particularly more so than is actually possible in everyday life." Binns further advised his readers that "radio is still a young science, and some of the most remarkable advances in it have been contributed by amateurs—that is, by boy experimenters. . . . Don't be discouraged because Edison came before you. There is still plenty of opportunity for you to become a new Edison, and no science offers the possibilities in this respect as does radio communication."[47]

Though such promises of fame and adventure may have been overenthusiastic, these popular accounts were simply embellishing reality. A young man's life was indeed made more exciting by involvement in radio. The amateurs came to feel that their lives were intertwined with truly significant events as they overheard messages about shipwrecks or political developments and transmitted these messages to others. Those amateurs who heard Jack Binns's distress signals became celebrities by association. One remembered that "the few boys in school in the area who claimed to have received the distress call were local heroes for a time, and they made a number of converts to the radio amateur hobby among the more technically minded youngsters."[48] Hearing "the news" first, the night before other Americans would read the story in the newspapers, imbued the amateur with an aura of being "in the know." As Francis A. Collins wrote in *The Wireless Man*, "Over and over again it has happened that an exciting piece of news has been read by this great audience of wireless boys,

Figure 2
Publicity photograph of Jack Binns.

long before the country has heard the news from the papers. . . . a wide-awake amateur often finds himself independent of such slow-going methods of spreading the news as newspapers or even bulletin boards."[49] The amateurs could feel part of an inner circle of informed people because they heard the news as it happened and because they were tapping point-to-point messages meant only for certain ears, not broadcasts intended for just anyone.

Equally important was the novelty of contacting strangers across the miles. Although many operators hoped to hear dots and dashes coming from thousands of miles away, making contact over a distance as small as 10 or 15 miles was reportedly a "thrilling experience."[50] In a culture that was becoming more urbanized and in which social networks were becoming increasingly fragmented, many strangers became friends through wireless. The fraternity that emerged possessed the fellowship felt among pioneers. These young men were exploring and comparing their findings on relatively uncharted and mysterious territory. As one amateur explained, "The eagerness and frankness in distributing the results of our findings undoubtedly molded the form of fellowship which is such a striking quality of the amateurs."[51]

An unusual social phenomenon was emerging. A large radio audience was taking shape that, in its attitude and its involvement, was unlike traditionally passive audiences. Collins summarized the development in *The Wireless Man*: "An audience of a hundred thousand boys all over the United States may be addressed almost every evening by wireless telegraph. Beyond doubt this is the largest audience in the world. No football or baseball crowd, no convention or conference, compares with it in size, nor gives closer attention to the business at hand."[52] This was an active, committed, and participatory audience. Out of this camaraderie emerged more formal fraternities, the wireless clubs, which were organized all over America. By 1912, the *New York Times* estimated that 122 wireless clubs existed in America.[53] Most of the club meetings took place "in the air" on a prearranged wavelength. The chairman called the meeting to order by sending out his call letters, and the members signified their attendance by answering with their own. During these meetings, the amateurs usually shared technical problems and solutions and drilled each other on transmission skills. A Chicago wireless club broadcast a "program" every evening "as a matter of practice for amateur operators in receiving." "The bulletin usually consisted of an article of some electrical or telegraphic interest. . . . sometimes the program was varied by sending passages in foreign languages, to quicken the receiving ears of the amateur operators."[54]

Gradually, an informal wireless network was established as the different clubs relayed messages for each other to points too far to reach with most amateur sets. "Message handling—for pleasure, for friends, in time of emergency—was rapidly becoming the predominant theme in amateur radio."[55] In March 1913, the midwest was hit by a severe windstorm that blew down the telegraph and telephone lines. The local amateurs handled the region's communications by relaying messages in and out. Such impromptu public-service gestures led some amateurs to advocate the establishment of better-organized communication among operators. One radio enthusiast, Hiram Percy Maxim, believed that the amateurs needed a national organization to establish a formal relay system or network to serve all amateurs. Through his Hartford Radio Club, he sent out invitations in March 1914 to amateurs to join a league and have their stations become official relay stations. The name of his organization was the American Radio Relay League (ARRL), and response to his invitations was so enthusiastic that within four months the league boasted 200 official relay stations across the United States.[56] Thus, in 1914, a grassroots, coast-to-coast communications network existed. Upon the formation of the ARRL, *Popular Mechanics* noted: "The coming of wireless telegraphy has made it possible for the private citizen to communicate across great distances without the aid of either the government or a corporation, so that the organization of the relay league actually marks the beginning of a new epoch in the interchange of information and the transmission of messages."[57]

People were becoming invisibly bound together by and in the airwaves, not by necessity, but for fun, to learn, and to establish contact with others. Those involved in the new hobby saw larger-than-life reflections of themselves in popular books, magazines, and newspapers. We cannot tell whether popular culture helped increase participation, but we do know, from the reminiscences of amateurs alive at the time, that the popular culture articulated the hopes and dreams invested in wireless. The technology gave the amateurs the means to communicate without wires. The popular culture sustained their visions of being on the cutting edge of technological progress. By now, the amateurs were not so much envisioning the future as they were laying the groundwork for it.

What did the amateurs' ever-increasing activity in the airwaves portend? The emergence of the amateurs and their often unrestrained fervor influenced both the immediate and the long-range regulatory, technical, and social developments in broadcasting. As increasing numbers of amateurs took wireless communication into their own hands, their activities became a nuisance to wireless companies and the government. In contrast with the early visions, which suggested there would be room for all in "the air," it was sadly discovered that the spectrum could accommodate only so many transmitters in a given area, especially when many of the sending stations emitted highly damped waves from crude spark gaps. Experimenters also learned that point-to-point, directional signaling was, at the time, impossible to achieve. And while some amateurs were skilled operators, devoted to serious experimentation, others were novices who clogged the airwaves with inconsequential and slowly sent messages. Francis Hart, a wireless operator in New York City from 1907 to 1911, described the congestion in his log book: "The different kids around here raise an awful noise, all try to talk at once, call when anybody is in and never use any sense, half can't read 4 words a minute and sit calling everybody within 20 miles and can't hear 800 feet from another station." He commented on one amateur's conversations: "FH is a very good reader, but he tries to say too much at one time then the poor reader makes him repeat it and they keep that blooming business up for hours."[58] As this sort of interference increased, so did "malicious" interference, which began to give the amateurs a bad reputation. Posing as military officials, some amateurs dispatched naval ships on fabricated missions. Navy operators received emergency messages warning them that a ship was sinking off the coast. After hours of searching in vain, receivers heard the truth: The supposedly foundering ship had just arrived safely in port.[59] Navy operators at the Newport Naval Yard complained that amateurs sent them profane messages. Others reportedly argued with Navy operators over right-of-way in the air.[60] *The Outlook* reported that during what Navy operators claimed was an emergency situation, amateurs refused to clear the air, "some of the amateurs even arguing with the Navy men over the ownership of the ether."[61] In another instance, when a Boston amateur was told by a naval operator to "butt out," he reportedly made the following classic remark: "Say, you Navy people think you own the ether. Who ever heard of the Navy anyway? Beat it, you, beat it."[62]

What had developed was the inevitable situation of too many people wanting access to the airwaves at the same time, with no guidelines for establishing priority. Too many people had embraced the invention and its possibilities. During this era, before 1912, no spectrum allocation had occurred, and all operators—amateur, commercial, and naval—vied with each other for hegemony. Military lobbyists in Washington, citing safety at sea and national security as reasons, advocated legislation that would ban amateurs from transmitting over the then-preferred portion of the spectrum.

The amateurs could not accept that the Navy should suddenly step in and claim the airwaves for itself in the name of national security when the Navy had done little to develop or refine wireless. The amateurs asserted that they had much if not more right to transmit, because they had worked and experimented to earn that right. While the Navy relied on outdated apparatus and unskilled or disinterested operators, the amateurs claimed to promote technical progress and individual commitment. The airwaves were a national resource, a newly discovered environment, and the amateurs claimed that their early enthusiasm and their technical work had entitled them to a sizable portion of the territory. They asked where the Navy had been when they were translating vision into practice. Much as the nineteenth-century pioneers had obtained squatters' rights by cultivating the land on which they had settled, the amateurs had developed a proprietary attitude toward the airwaves they had been working for the past five years. They granted that there were a few outlaws in their midst, but they argued that the alleged violations did not justify the exclusion of all individual operators by the government.

Ultimately, the amateurs lost. During the *Titanic* disaster of April 1912, interference from amateur stations trying to relay as well as elicit news was so great that within four months the Congress banished their transmissions to a portion of the spectrum then deemed useless: short waves. The Radio Act of 1912 also required that amateurs be licensed, and imposed fines for "malicious interference."[63]

What impact, then, could a group of operators have on radio broadcasting who were, by 1912, banished to this etheric reservation? One scholar who has studied how Americans have managed the airwaves points out that "relatively deprived users" were "virtually forced to innovate spectrum-economizing, spectrum-developing technology."[64] It was the amateurs, the recently deprived users, who would pioneer one of the biggest breakthroughs in radio: short-wave broadcasting. One of the more famous amateurs, Edwin Armstrong, developed the regenerative or feedback circuit, which amplified the often feeble signals coming in over the receiving antenna. Thus, the amateurs' technical contributions remained signifiant. Also, less creative amateurs, by reporting results to others, provided the more serious experimenters with valuable data on performance and results.

In the years after the Radio Act of 1912, the amateurs not only advanced radio technology but also anticipated broadcasting. Between 1910 and 1920, amateur stations began to broadcast music, speech, and even advertising. By 1917, amateurs were relaying messages not just regionally but from coast to coast, demonstrating the benefits of a national communications network. Some of the early amateur stations became commercial stations in the 1920s. Frank Conrad of Pittsburgh, a radio amateur and a Westinghouse employee, operated in his garage a small amateur station that is usually credited with inaugurating the broadcasting boom. In 1920, Conrad's station was moved to a portion of the Westinghouse plant and became KDKA. It was the amateurs who demonstrated that, in an increasingly atomized and impersonal society, the nascent broadcast audience was waiting to be brought together. Using the airwaves to inform, entertain, and connect the general public was, before 1920, still not in the corporate imagination. Institutions continued to view radio as merely a substitute for cables, a technology that would provide long-distance, point-to-point communications.[65] The Radio Corporation of America, which was formed to establish an American-controlled international network, was compelled to reconsider its purpose and its goals shortly after its formation in 1919. The industry that would come to control radio broadcasting by the late 1920s had to respond to a way of using the airwaves pioneered by the amateurs.

In their fight to retain access to the preferred portion of the spectrum in 1912, the amateurs claimed to be surrogates for "the people," who, they declared, were the rightful heirs to the spectrum. In congressional testimony and letters to magazines and newspapers, the amateurs insisted that individuals, not the Navy or big business, should determine how the airwaves were used.[66] This democratic ideology, manifested both in rhetoric and in practice throughout the teens and the twenties, contributed to the legitimation of the public's claim to and stake in the air. The Communications Act of 1934, which established the Federal Communications Commission and required the licensing of all radio stations, mandated that these stations serve "the public interest, convenience, and necessity."

The turn-of-the-century predictions about radio's future applications had not come true. They had been based on a misunderstanding of how the invention worked, and they assumed that radio, by itself, could change the world. Yet even dreams that do not come true can have an effect. By encouraging and romanticizing the amateurs' hobby, these visions fostered experimentation among members of a subculture who had neither a corporate nor a political agenda. The predictions also articulated and reinforced the belief that this technology could and should be accessible to the greatest possible number of Americans.

Such dreams did not die in the 1920s. They were simply transformed to accommodate the new reality of institutional management. Radio was now indeed firmly embedded in a corporate grid, and the new visions of the 1920s, while still very enthusiastic, made concessions to this centralized control. Just as Frank Conrad's radio station moved from his garage to Westinghouse, so too did visions of radio's uses and benefits begin to reflect corporate agendas. In the 1920s there was little mention of world peace or of anyone's ability to track down a long-lost friend or relative halfway around the world. In fact, there were not many thousands of message senders, only a few. The theme of isolation was still central, but instead of the separation of one individual from another the predictions of the 1920s focused on certain individuals' separation from the mainstream of American culture. "All isolation can be destroyed," proclaimed Stanley Frost in a 1922 article entitled "Radio Dreams That Can Come True." Now radio had the potential to be a "tremendous civilizer" that would "spread culture everywhere" and bring "mutual understanding to all sections of the country, unifying our thoughts, ideals, and purposes, making us a strong and well-knit people." This audience would be passive: "We do not even have to get up and leave the place," exclaimed a *Collier's* contributor. "All we have to do is to press a button."[67] Thus, through radio, Americans would not transcend the present or circumvent corporate networks. In fact they would be more closely tied to both. Visions and reality were merging.

Yet at least one vision of how radio would bring about a utopian future persisted. In the spring of 1919, Marconi announced that several of his radio stations were picking up very strong signals "seeming to come from beyond the earth."[68] Nikola Tesla, another prominent inventor, believed these signals were coming from Mars. Articles in newspapers and magazines speculated about the signals and, reiterating Preece's prediction of 20 years before, asked "Can we radio a message to Mars?" *Illustrated World*, a magazine that popularized recent technical developments, urged that America try; only then would the Martians know that "their signals were being responded to, and that intelligent beings actually inhabit the earth."[69] The writer then added: "We can imagine what excitement this would cause on Mars." But the most important reason for trying to contact Mars was to learn what it was assumed they must know about improving, even perfecting, the quality of life. As *Illustrated World* put it, "It is not unreasonable to believe that the whole trend of our thoughts and civilization might change for the better." Martians would not only view our civilization with considerable detachment, but they would also presumably give us all the secret answers. Once again, through radio, we might be able to escape the institutions in which we found ourselves ensnared.

The idea of sending radio signals to Mars was in many ways the most revealing and poignant of the visions surrounding radio. It exposed a sense of isolation, insecurity, and dissatisfaction over things as they were. It revealed that, despite the failure of past predictions, many Americans were still inclined to view certain technologies as autonomous, as possessing superhuman or magical powers. This willingness to invest certain inventions with individual hopes and cultural aspirations has permitted the corporate sphere to exploit a range of technologies to profitable ends, but it has also led certain Americans, such as the amateur operators, to take technology into their own hands and, in the process, profoundly influence the course of technical change.

Notes

1. *Review of Reviews* 18 (December 1898), p. 715.

2. "A Century of Progress in the United States," *Scientific American* 83, no. 26 (1900), pp. 400–402.

3. *Popular Science Monthly*, December 1897, p. 263.

4. Carl Snyder, "Wireless Telegraphy and Signor Marconi's Triumph," *Review of Reviews* 25 (February 1902), p. 173; *New York Times*, January 14, 1902, p. 1; January 15, 1902, p. 8.

5. P. T. McGrath, "The Future of Wireless Telegraphy," *North American Review* 175 (August 1902), p. 274.

6. *Current Literature* 35 (November 1903), p. 533.

7. Ibid.

8. Ray Stannard Baker, "Marconi's Achievement," *McClure's Magazine* 18, no. 4 (1902), p. 294.

9. Georges Iles, "Marconi's Triumph," *World's Work* 3 (February 1902), p. 1785.

10. Ibid.

11. P. T. McGrath, "Authoritative Account of Marconi's Work in Wireless Telegraphy," *Century Magazine* 63 (March 1902), pp. 779–780.

12. *Popular Science Monthly* 61, no. 1 (1899), p. 72.

13. *New York Times*, August 15, 1899, p. 6.

14. John Hall Ingham, "Wireless Telegraphy," *Atlantic Monthly* 85 (January 1900), p. 137.

15. *New York Times*, May 7, 1899, p. 20.

16. McGrath, "Authoritative Account of Marconi's Work," p. 782.

17. "New Wonders with 'Wireless'—and By a Boy," *New York Times*, November 3, 1907, pt. 5, p. 1; H. D. Jones, "Pocket Wireless," *World Today* 20 (June 1911), p. 747; "Pocket Wireless," *Literary Digest* 48 (January 31, 1914), p. 201.

18. See Michael Schudson, *Discovering the News* (New York: Basic Books, 1978); Dan Schiller, *Objectivity and the News* (Philadelphia: University of Pennsylvania Press, 1981).

19. *New York Times*, August 15, 1899, p. 6.

20. Ibid.; *New York Times*, March 9, 1902, p. 6; May 1, 1902, p. 8; May 7, 1899, p. 20.

21. Snyder, "Wireless Telegraphy," p. 173.

22. See Frank Fayant, "The Wireless Telegraph Bubble," *Success Magazine* 10, no. 157 (1907), p. 387; Susan J. Douglas, Exploring Pathways in the Ether: The Formative Years of Radio in America, Ph.D. diss., Brown University, 1979, pp. 259–272.

23. Robert A. Morton, "The Amateur Wireless Operator," *The Outlook* 94 (January 15, 1910), p. 131.

24. *Electrical World* 51, no. 9 (1908), p. 423; 54, no. 24 (1909), p. 1401.

25. For information on crystal receivers see History of the Wireless Specialty Apparatus Company, dictated by G. W. Pickard to G. H. Clark (unpublished, 1931); "Radioana," *Electrical World* 48, no. 23 (1906), p. 1100; A. P. Morgan, *Wireless Telegraphy and Telephony* (New York: Norman W. Henley, 1912), pp. 52, 57; Elmer E. Bucher, *Practical Wireless Telegraphy* (New York: Wireless Press, 1917), p. 132; Lee De Forest, *Electrical World* 48, no. 10 (1906), p. 491; WSA to Director of Naval Intelligence, London, July 10, 1908, "Radioana," *Electrical World* 48, no. 21 (1906), p. 994.

26. Allen Chapman, *The Radio Boys' First Wireless* (New York: Grosset and Dunlap, 1922), p. 63.

27. Information on Model T coils provided by Hugh G. J. Aitken.

28. Donald G. Little, reminiscences (unpublished manuscript), p. 5.

29. E. L. Bragdon, reminiscences (unpublished manuscript, Columbia Oral History Library), p. 4.

30. *New York Times*, May 31, 1909, p. 1.

31. Edgar S. Love, reminiscences (unpublished manuscript, Columbia Oral History Library), p. 2.

32. Clinton B. De Soto, *Two Hundred Meters and Down, The Story of Amateur Radio* (West Hartford, Conn.: American Radio Relay League, 1936), p. 3.

33. *Electrical World* 57, no. 13 (1911), p. 760.

34. *New York Times*, March 29, 1912, p. 12.

35. "New Wonders with 'Wireless'—and By a Boy" (note 17).

36. Francis A. Collins, *The Wireless Man* (New York: Century, 1912), p. 29.

37. Allen Chapman, *The Radio Boys with the Iceberg Patrol* (New York: Grosset and Dunlap, 1924).

38. Not only is there an increase in the number of articles about wireless (and then radio), but also there is an increase in articles predicting what radio will be used for in the future and in how-to articles for amateurs.

39. Chapman, *The Radio Boys and the Iceberg Patrol*, pp. 26–27.

40. Victor Appleton, *Tom Swift and His Wireless Message* (New York: Grosset and Dunlap, 1911), pp. 179–196.

41. Chapman, *The Radio Boys and the Iceberg Patrol*, p. 23.

42. "New Wonders with 'Wireless'—and By a Boy" (note 17).

43. *New York Times*, January 31, 1909, p. 18; April 29, 1910, p. 18.

44. Hugo Gernsback, letter to the editor, *New York Times*, March 29, 1912, p. 12.

45. Lloyd Jacquet, "The Heritage of the Radio Club of America," in Fiftieth Anniversary Golden Yearbook (New York, 1959), p. 4.

46. *Current Literature* 46, no. 3 (1909), p. 248.

47. Jack Binns's forewords to Chapman's *The Radio Boys' First Wireless* and *The Radio Boys with the Iceberg Patrol.*

48. Edgar Felix, reminiscences (unpublished manuscript, Columbia Oral History Library), p. 4.

49. Collins, *Wireless Man*, pp. 29–31.

50. Love (note 31), p. 5; Bragdon (note 29).

51. Jacquet, "Heritage."

52. Collins, *Wireless Man*, p. 26.

53. *New York Times*, January 31, 1909, p. 18; March 29, 1912, p. 12.

54. *Electrical World* 56, no. 3 (1910), p. 139; Collins, *Wireless Man*, pp. 42–47.

55. De Soto, *Two Hundred Meters*, pp. 37–41.

56. Ibid.

57. Ibid., p. 40.

58. Francis Hart, log book entry of November 23, 1907, Radioana Collection, Smithsonian Institution, Washington, D.C.

59. *New York Times*, January 28, 1910, p. 8; January 30, 1910, p. 4.

60. *Electrical World* 57, no. 13 (1911), p. 760; Morton, "Amateur Wireless Operator."

61. Morton, "Amateur Wireless Operator," pp. 132–133.

62. Ibid.

63. For accounts of amateur interference and public reaction, see *New York Times*, April 10–23, 1912; "*Titanic* Disaster: Hearing Before a Sub-committee of the Committee on Commerce, United States Senate," Senate Documents, 62nd Congress, 2nd session, December 4, 1911–August 26, 1912, vol. 28, document 726.

64. Harvey J. Levin, *The Invisible Resource* (published for Resources for the Future, Inc., by Johns Hopkins University Press, Baltimore, 1971), pp. 9, 18.

65. See Gleason Archer, *History of Radio to 1926* (New York: American Historical Society, 1938); Rupert W. MacLaurin, *Invention and Innovation in the Radio Industry* (New York: Macmillan, 1949).

66. "Hearings Before a Subcommittee of the Committee on Naval Affairs of the House of Representatives on H. J. Resolution 95" (Washington, D.C.: GPO, 1910); *New York Times*, April 29, 1910, p. 18; De Soto, *Two Hundred Meters*, p. 32.

67. Stanley Frost, "Radio Dreams That Can Come True," *Collier's* 69 (June 10, 1922), pp. 9, 18.

68. "That Prospective Communication with Another Planet," *Current Opinion* 66 (March 1919), p. 170. See also "Those Martian Radio Signals," *Scientific American* 122 (February 14, 1920), p. 156.

69. Thomas Waller, "Can We Radio a Message to Mars?" *Illustrated World* 33 (April 1920), p. 242.

3

Wasn't the Future of Nuclear Energy Wonderful?

Stephen L. Del Sesto

In his 1909 book *The Interpretation of Radium and the Structure of the Atom*, Frederick Soddy predicted that scientists would eventually learn to release vast amounts of energy from uranium. According to Soddy, "the energy of a ton of uranium would be sufficient to light London for a year. The store of energy in uranium would be worth a thousand times as much as the uranium itself, if only it were under control and could be harnessed to do the world's work in the same way as the energy in coal has been harnessed and controlled."[1] Nearly a half-century later, in a 1955 issue of the *Ladies' Home Journal*, it was proclaimed that in the near future nuclear energy would create a world "in which there is no disease . . . where hunger is unknown . . . where food never rots and crops never spoil . . . where 'dirt' is an old-fashioned word, and routine household tasks are just a matter of pushing a few buttons . . . a world where no one stokes a furnace or curses the smog, where the air is everywhere as fresh as on a mountain top and the breeze from a factory as sweet as from a rose." "Imagine," the article continued, "the world of the future . . . the world that nuclear energy can create for us." These were not the thoughts of an overactive imagination or the wild dreams of an obscure science-fiction writer, but those of Harold E. Stassen, President Eisenhower's Special Assistant on Disarmament.[2]

One is inclined to dismiss Soddy's remarks as early speculation in an area where few hard facts existed. In 1909 much of the basic science still needed to be undertaken and practical applications of nuclear energy were years away. But Stassen had the benefit of the knowledge gained in the ten years since the dawn of the nuclear age. Many people, in fact, shared his optimism about the future of atomic energy. Indeed, by the mid twentieth century nuclear utopianism became a significant strand of popular culture, drawing strength from the general optimism kindled by decades of scientific and technological progress.[3]

Several factors combined to produce utopian speculation about the uses of nuclear energy. First, throughout the U.S. government, which had taken the control of atomic energy out of the hands of the military in favor of civilian management with the Atomic Energy Act of 1946, it was believed that civilian uses should be demonstrated as rapidly as possible in order to downplay the association of all things nuclear with the bomb. This set officials to formulating (with the help of technical advisors) public policies aimed at encouraging feasible civilian applications. Today's civilian nuclear power program is only one of the results of this effort.[4] Second, scientists had recognized for many years that, at least in theory, the harnessing of nuclear energy could offer huge amounts of cheap power to industrial societies that relied upon continuous supplies of heat and energy to turn machinery in factories, run blast furnaces, heat homes, and propel various kinds of vehicles. Too readily, as it turned out, theoretical possibilities were converted to beliefs in imminent practical realities. Third, in the 1930s and the 1940s the public was fascinated with the powers of science and technology. New developments rapidly became news items of a sort that the press was glad to report to an eager public. Science stories sold copies, and it is not surprising that more than one reporter embellished the facts about developments in nuclear power.

These interrelated forces—government attempts to encourage and demonstrate civilian applications, the theoretical possibilities and the economic potential of harnessing the atom, and the public fascination with science and technology—resulted in countless visions of the new world of nuclear energy. And these forces reinforced one another, making the dreams appear more plausible and closer at hand than they really were.

These utopian visions essentially concerned activities that required large amounts of energy: industrial processes (including manufacturing, electricity generation, and the production of new chemicals and metals), transportation, earth moving, water production, and agriculture. In addition, many believed that nuclear energy would have important applications in medicine and public health, particularly in aiding medical research and improving radiation therapy. In most instances, all these applications were woven together into a larger vision of a new era of industrial civilization made possible by the almost limitless energy that nuclear fission (and fusion) would provide at relatively little cost. Indeed,

Figure 1
Life in the atomic age, as envisioned in *Collier's*, July 6, 1940.

perhaps the most common belief centered on the notion of "power too cheap to meter."

The new power source was envisioned as taking many forms, from giant central power stations to the "package power plant"—a mobile unit that could be rolled into place in the event of a disaster or a natural catastrophe. R. M. Langer, a research scientist turned popular science writer, wrote in 1941 that miniature nuclear power reactors "the size of a typewriter" would provide enough energy to power every home and factory in America without distribution lines, and that electricity would cost "less than one-tenth of a cent per kilowatt hour." He predicted a time of "universal comfort, particularly free transportation, and unlimited supplies of materials." Langer foresaw a future in which "Energy has become so cheap that it isn't worth making a charge for it. It is so convenient that there are no distribution costs. This means that freight as well as passenger transportation are public utilities; like the heat and light and water in your house, you don't have to pay for them at all."[5] Similarly, though perhaps with less enthusiasm, John J. O'Neill and Harry M. Davis envisioned the day when tiny fuel pellets would be used to make steam in large power plants.[6] Popular science magazines such as *Science Illustrated* and *Science News Letter* heralded the atomic age as one of new industrial processes and materials.[7] Useful new elements would be discovered, air pollution would cease, and there would be a bounty of cheap new goods and services. The limitless power of nuclear fission would unroll "the greatest magic carpet of all ages," wrote William McDermott in *Popular Mechanics* in 1945.[8] It seemed a new day had arrived.

By the time the Atomic Energy Act was signed into law, most of the members of the Congress and the federal government were convinced that peaceful nuclear energy should receive high priority as a national goal. In fact, section 1 of the Atomic Energy Act proclaimed the following: "It is reasonable to anticipate . . . that tapping this new source of energy will cause profound changes in our present way of life. Accordingly, it is hereby declared to be the policy of the people of the United States that the development and utilization of atomic energy shall be directed toward improving the standard of living, strengthening free competition among private enterprises so far as practicable, and cementing world peace."[9] This declaration of policy developed largely out of the views of the participants at the legislative hearings on the act. One participant, Robert Hutchins, the chancellor of the University of Chicago, called nuclear energy the greatest invention since the discovery of fire. He told the Congress that it promised to transform American society as dramatically as had electrification in the period after 1900.[10] Meanwhile, social scientists were also touting the revolutionary possibilities of nuclear power. Just as the legislative hearings were taking place in Washington, William Ogburn and Feliks Gross were writing in sociology journals that peaceful nuclear power would reduce or eliminate manual labor and increase leisure time.[11] The political scientist Charles E. Merriam wrote in the *American Journal of Sociology* that nuclear power promised "the greatest future ever spread before mankind with dazzling possibilities of life, liberty, and the pursuit of happiness."[12]

The President of the American Chemical Society, Dr. Harry M. Fisher, wrote in *American Magazine* in 1954 that experts in the atomic field believed, as he did, that "the world is not going to destroy itself with atomic bombs, but men will use the wonders of the atom to build a richer, fuller life for themselves and their families."[13] Ocean liners would be propelled by nuclear power plants, and "atomic bat-

Figure 2
The Nucleon, a model created by the Ford Motor Company Advanced Styling Department in 1958. According to a press release, the car would be powered by a replaceable, rechargeable nuclear reactor. Ford Motor Company.

teries" would power automobiles, washing machines, and "even tiny wrist-watch radios." Cheap, unlimited power produced from atomic fission would probably replace fossil fuels for industrial applications, with dramatic impacts on industrial communities: "Atomic power will transform the appearance of your home town. If you live in a community darkened by grime and afflicted with smog from power plants or factory smokestacks, you can look forward to seeing your town transformed into a clean, healthful place. Atomic furnaces, unlike coal furnaces, need no smokestacks."[14] In short, peaceful nuclear energy would transform the industrial processes and make the world a better place to live.

Next to forecasts of power "too cheap to meter," predictions of atomic-powered transportation were probably the most frequently voiced. Journalists heralded a revolution in both surface and air travel. Nuclear fission, they argued, could power everything from rocket ships to trains to automobiles.[15] The family car would soon have an "atomic engine" that would run on the immense amounts of energy contained in a tiny quantity of uranium or some other fissionable material. With nearly limitless fuel, "men would be free to tour the country in their Nuclear-8 sedans," Louis Cassels wrote in 1950 in *Harper's Magazine.*[16] Some ten years earlier in the same publication, Pulitzer Prize winning science journalist John J. O'Neill had written that in a nuclear car "one might be able to travel as many thousand miles as he wished without having to bother with fuel bills," and that oil consumption and maintenance would be insignificant because the gasoline engine would be replaced by lighter, more efficient steam-cycle atomic engines with nearly inexhaustible power packs.[17] S. C. Gilfillan, a research sociologist at the University of Chicago studying the social effects of technological innovation, claimed in 1945 that atomic-powered vehicles would be less expensive to maintain and fuel, would have more speed and efficiency, and would possess improved cargo-carrying capacity. Atomic engines would be smaller and lighter and would not need the cumbersome structures of conventionally powered vehicles.[18] The idea that cars and trucks would be fueled for life and run by extremely efficient atomic engines was very popular throughout the 1940s.

Perhaps the most fantastic visions of the nuclear car were those put forth by R. M. Langer in *Collier's* in 1940 and in *Popular Mechanics* in 1941. Langer predicted that cars would be constructed like airplanes, with an external skeleton of plastic and other light materials that would be "transparent where necessary." Such cars would have neither batteries, transmissions, nor heavy differentials, but would be light and fast, powered by a "butterfly" motor—a "U-235 engine capable of any speed or torque" and whose qualitites would include "ease of manufacture, simplicity of operation and control, versatility of performance and portability." (The butterfly motor was to be, basically, a high-pressure steam engine turning turbine wheels welded to a single shaft.) Langer described this wonderful car further:

> Under the body is a water tank that helps shield the occupants from radiations emitted by the uranium and also puts the center of gravity low. Each wheel hub has on it a very small reversible turbine motor. . . . This car needs no differential or clutch. The uranium must be detachable since it probably will outlast the car. Recovering its own exhaust steam, such a car could travel without stopping until it wore out its tires or needed other servicing. Using the newly developed vertical propellers, such a vehicle might be converted into a high-speed, long-range vessel to travel on water, and might be furnished with unfolding wings for travel through the air.[20]

The atomic car would be comfortable and spacious, having one or more "rooms" or "compartments" with overhead windows and blinds. The driver would be able to operate the car easily with a hand-held remote control device that would free him to do other things while traveling in ease and comfort. Describing the car in *Collier's*, Langer wrote: "The wheels are big as tractor wheels, to minimize disturbance because of bumps and to prevent damage to lawns and fields. Roads are practically unnecessary, except for main thoroughfares used mostly for freight transportation. There is no conceivable use for a railroad."[21] Moreover, the car would be suspended from an overhead axle, which would allow easy banking on turns and would permit occupants to "write letters or do chores while under way."[22] On the highway, cars would be "kept in lanes by buried pilot cables similar to the electronic devices used to guide ships into harbors."[23] What highways were still needed would be "paved" by an atomic device—as Langer put it, "with intense heat available a road-building machine would be able to fuse all the dirt in its path into lava, making in one operation a wide rock highway ideal for smooth travel."[24]

Of course, no nuclear car ever covered a single mile, on or off the highway. In fact, there is little evidence that the idea of such a vehicle ever progressed much beyond the wild speculations of Langer and others. E. V. Murphree, Vice-President of Research and Development at Standard Oil of New Jersey and a participant in the Manhattan Project, rejected all such speculation. In a 1946 article in *Popular Mechanics*, he wrote: "The popular fancy of pellets of uranium burned like oil or coal in something like an internal combustion engine does not fit present facts."[25] One reason that atomic engines would not be used in power autos, Murphree noted, was that they would require heavy and cumbersome shielding materials to protect the occupants. The nuclear car was simply out of the question. Similarly, *Science News Letter* quoted Edward U. Condon, then Director of the National Bureau of Standards, as saying that nuclear power was unlikely to be used in cars, planes, or locomotives; there were just too many engineering difficulties, and nuclear engines would be too heavy, expensive, and inefficient for such vehicles. For large ships and central power plants, Condon believed, nuclear power would probably work.[26]

Other enthusiasts looked to a future in which nuclear energy would propel airplanes and rocket ships. In 1960, in a story about the "magic of atomic power," *Newsweek* predicted that in a few years the skies would teem with large commercial airliners powered by nuclear energy.[27] Such dreams were even shared by some of the more specialized professional aviation magazines, such as *Aviation Weekly* and *Flying*. The latter magazine ran a story in 1957 in which the author claimed that atomic planes would "revolutionize flight with their unlimited range and endurance." "On less than one pound of enriched uranium," it was predicted, "an airplane will be able to fly 100,000 miles."[28]

This fascination with nuclear-powered flight was not new, however. In 1940, R. M. Langer had prophesied that airplanes would be powered by "the ejection of high speed particles," a "means of propulsion in the same manner that the ejection of water can be made to cause a lawn sprinkler to rotate."[29] Langer went on to say:

> Such a device would eliminate the necessity for a propeller, and would do well what a helicopter can do only badly. It would overcome gravity, and cause an object to rise vertically. Airplanes of this character would be able to fly at any height above the earth because they are not dependent upon the atmosphere to keep them aloft. At a height of, say, fifty miles, the resistance of the air is so slight that such craft could attain speed of several thousand miles per hour. You could have supper in Paris, and speed across to New York faster than the sun, to see a matinee performance on Broadway. You can see what happens to time; and, of course, geographical and national boundaries will lose their meanings too.

In 1948 both *Time* and *Newsweek* ran stories quoting Andrew Kalitinsky of Fairchild Engine and Airplane Corporation, the chief engineer working with the Atomic Energy Commission on the nuclear plane. While Kalitinsky was cautious in his statements and stressed the complexities of nuclear flight, *Newsweek* indicated that he thought the "ultimate deployment of atomic aircraft was certain."[30] This corroborated a *Science News Letter* report of April 1946 that attributed to Glenn Seaborg, a University of California scientist and the co-discoverer of plutonium, the opinion that airliners could be powered by U-235 or plutonium if the government could promulgate successful international controls to prevent diversions of fissionable materials for the manufacture of bombs. According to the article, Seaborg said that "after the atomic pile has been freed of the load of graphite now necessary for keeping the output of energy within safe bounds, the atomic energy unit will sprout great wings and take to the upper air."[31] Although Seaborg seems to have been carried away by his metaphor, he was not alone in thinking that atomic power would allow stratospheric travel at great speed and convenience.[32] One writer predicted that experimental nuclear-powered aircraft would take to the air as early as 1950.[33] "The first test aircraft," he wrote, "will probably be jet-propelled, will carry no crew, and will be controlled from the ground or from other planes."

The speculation about nuclear-powered rocket ships was much the same. Many fantastic, grandiose claims were made about the possibility of interplanetary exploration and travel. Even some distinguished scientists conceded the possibility of nuclear rocket ships. Alvin Weinberg, a veteran of the Manhattan Project, told *The New Republic*: "Space travel, which has until recently been deemed fantastic, must now be considered more seriously."[34] (Of course, Weinberg's comment was only a passing remark and can scarcely be considered fantastic in the strict sense of the word.) S. C. Gilfillan said that nuclear rocket propulsion could probably be achieved by vaporizing, through atomic heat, "some cheap liquid substance such as water," and that trips around the moon would thus be possible in the not-too-distant future.[35] Pulitzer Prize winning physicist William Laurence suggested the probability of sending men to the moon and Mars, and predicted that atomic rockets would permit easy exploration of the solar system.[36] The head of the United States Rocket Society told *Popular Mechanics* writer William McDermott in 1945 that "a trip to the moon by way of atomic power is not only possible, but one may before long go joyriding among the planets."[37] All that was needed to achieve these herculean feats was "the mobilization of the best brains in humanity," according to another influential author.[38]

Such unrestrained faith in organized science and technology was a common theme of many of the early sensationalizers. However, R. M. Langer, in his usual fashion, seemed to have the details of nuclear rocket travel "worked out" better than anyone else. He believed that atomic-powered rockets would be attached to conventional airplanes. Indeed, he had a large store of visionary ideas about the future of aeronautics:

> Jet propulsion will free the airplane from earth. For both local low-altitude trips and long-range flights at high altitudes we will probably always use the simple airfoil or flying-wing design. We will use the principle of rocket power but there will be no need to employ the bullet shape of present rockets. We will fly several thousand miles per hour at several thousand feet above the surface and will have to slow down in descending to the surface to avoid being burned up like a meteor.
>
> The flying-wing driver will have a gyroscope, the main propulsion jets, and a few directable controls at his command. Compressed air or steam for the jets might do for landing or leaving the ground but for greater efficiency we might very well use steel vapor for propulsion at higher altitudes. Vapor from boiling steel would have powerful thrust and would condense as harmless dust.[39]

Of course, it is unlikely that many people believed these fantastic predictions about nuclear-powered rockets and jet planes, and many experts pointed out the flaws. The physicist Luis Alvarez, of the University of California at Berkeley, coolly refuted the atoms-for-flight notion. "There is no obvious or simple way in which to use atomic energy for space ships," he wrote, and to use atomic rockets would be "to do an easy thing the hard way."[40] Alvarez also argued that, though it was "*technically* feasible," airplanes would "probably *not* be propelled by atomic power in the next ten years."[41] Fuel, weight, and shielding problems seemed intractable, he said. Another technical expert, Leonard Katzin, argued: "Not only will it be impossible to have an indefinitely small nuclear power source based upon fission of uranium, but any such power source must be heavy and awkward, and quite dangerous. . . ."[42] In short, to many knowledgeable observers "power pills," "nuclear black boxes," and other "individual" atomic power units were mere fantasies.

Notwithstanding such criticisms, dreams of atomic aircraft lived on in many scientific, military, and governmental circles. The Air Force and the AEC jointly began the Aircraft Nuclear Propulsion (ANP) program in 1946, envisioning the military advantages of a long-range nuclear plane. Although some experts (including J. Robert Oppenheimer) opposed the ANP program on technical grounds from the beginning, sufficient official and corporate support existed to keep it alive for years. In 1959, the Joint Committee on Atomic Energy said that the development of a nuclear plane was in the national interest and that the United States had to achieve nuclear flight as soon as possible, not only to meet military requirements but also to boost world confidence in American scientific capabilities. In 1961, after expenditures of more than a billion dollars, the new president, John F. Kennedy, managed to kill the costly pursuit of nuclear-powered flight.[43]

Throughout the same period, officials held out great hope for nuclear rockets, and money was poured into research-and-development programs. In the spring of 1958 the 85th Congress passed an amendment to the Atomic Energy Act of 1954 establishing a Division of Outer Space Development within the AEC, made provisions for an Outer Space Advisory Committee, and created a National Laboratory for Outer Space Propulsion. An immediate $50 million was appropriated to carry out the provisions of the amendment.[44] Project Pluto, a joint AEC–Air Force study, was instituted at the request of the Department of Defense in 1956 in order to demonstrate the feasibility of nuclear ramjet propulsion. Between 1956 and 1958, after a number of feasibility studies, Project Pluto was reoriented toward supersonic low-altitude applications, partly in response to the encouraging results of the missile testing programs. Meanwhile, a second nuclear-rocket plan, Project Rover, had been underway since 1955 at the Los Alamos Scientific Laboratory. By 1959 two experimental reactors, Kiwi-A and Kiwi-A′, had been successfully ground-tested, and it looked as if work might begin on a prototype model. The chairman of the Joint Committee on Atomic Energy, Senator Clinton P. Anderson, was so convinced of the feasibility of nuclear rocket propulsion that he argued on the Senate floor on September 1, 1960, that the test data warranted full-scale policy formulation. He advocated an extensive ten-year program.[45]

Although enthusiasm for Senator Anderson's program was minimal, science journalists were still advocating nuclear propulsion for space travel. In the spring of 1961, Harold L. Davis, Associate Editor of *Nucleonics Magazine*, called nuclear propulsion the "key to a new technological era." He wrote: "As the steam engine gave man mastery of the sea and the gasoline engine gave him mastery of the air, so nuclear power is destined to make possible man's conquest of space. Nuclear power for space missions will mean the difference between instrumental probes and practical manned spacecraft."[46] Despite such optimistic claims, the nuclear rocket ship, like the nuclear plane, never left the ground. The dream of nuclear flight did not die easily, though. As the government scuttled plans for further research and development, some were still asking why such promising possibilities were being ignored.

There were yet other wonders to be unleashed by the nuclear genie. Nuclear energy would help establish complete control over the natural environment, claimed the prophets. Their schemes ranged from weather modification, to the control of photosynthesis in plants, to the use of nuclear explosives for excavation.

In July 1948, *Christian Century* boldly proclaimed that "atomic research may end the world's hunger." The editorial noted that recent experiments promised "a vast increase in the world's food supply within the next year or so!" It reported that scientists at the University of California at Berkeley were experimenting with atomic energy and algae, and that soon it would be possible to "produce all the proteins, fats, sugars, and vitamins all the people of the world need to have a full and even luxury diet at little cost."[47] Finally, the article predicted that experiments would make possible the use of seawater to irrigate arid regions, so that the world's hunger problems would truly be over.

The idea that nuclear energy would turn deserts into vast agricultural gardens was fairly popular. An article in *Science Illustrated* predicted in June 1946 that the nuclear power plants (for desalinization of sea water and production of electricity) would make the Sahara bloom. "Atomic engines" could be installed in isolated regions "where *no* other power source was available," wrote Louis Cassels in 1950.[48] Agriculture and economic development would allow new human settlements to flourish in wastelands and underdeveloped countries.[49]

The "real" agricultural revolution, however, would probably result from experiments in which "tagged atoms" would be used as tracers to explore the process of photosynthesis. In 1946, writing for the *Saturday Evening Post*, William Laurence put it this way:

> By learning how the plant builds up its food substance from carbon dioxide, water and a few minerals—we could also build these other substances out of tagged elements—we may learn to use the same substances and sunlight for the direct production of food. We would no longer be dependent on the soil to give us our daily bread. Man at last may be able to produce enough food to provide abundantly for the world's population. The nitrogen in the atmosphere, the water in the rivers and some of the common elements in the soil and in the sea will be the raw materials out of which he will "grow" his foods.[50]

Other writers interpreted the coming agricultural revolution in different ways. Writing in *Coronet* in 1948, Harold Wolff predicted that by 1950 scientists would perfect "atomic techniques" to make plant hormones that would "help more crops and kill more weeds."[51] The introduction of "radioactive fertilizer" at the right time "might reach the sex cells of the plants at the time they are forming new cells"; this would produce thousands of mutations and make it possible to grow "sturdier, more productive and cheaper crops," said Wolff.

Still others stressed the social and technological aspects of the nuclear agricultural revolution. *Science News Letter* reported in 1946 that a University of Chicago sociologist, Louis Wirth, believed that nuclear-based agricultural techniques would change the social patterns of agriculture throughout the world. Wirth believed that increased productivity would level social differences between rich and poor and equalize the standard of living worldwide.[52] Wirth's colleague at Chicago, S. C. Gilfillan, speculated in 1945: "Agriculture will be much helped by cheaper power and heat, including perhaps small engines to power a hand tool, like those mechanisms used in industry where there is an electrical connection. At the same time there will be less need for horses (and their fodder) and a great impetus to the end of agriculture as we know it. This will come by two routes, by synthesis and mineral substitutes for farm products, and by hydroponics or soilless agriculture."[53]

Other predictions were even more fantastic. One writer insisted that nuclear energy applied to agriculture would allow the manufacture of food, "on an industrial basis," from "water and limestone."[54] Just how this would be done, however, was not specified. According to Langer, nuclear energy might also allow plants and crops to be grown underground, with artificial light and heat, "using the technique of hydroponics, in the water that flows through the rocks just below [one's] underground ranch."[55] People would simply "proceed to their garden to satisfy their taste for fresh fruits which they pick from dust-free, sterile plants, and prepare in a few moments in the high-frequency cooker their favorite breakfast foods." Langer went on to explain how all this would be possible:

> . . . the big, sudden change brought by U-235 is that man is no longer dependent upon the sun for his cheapest and most essential manufacture—the production of food. This means that any country, with any climate, at any time of the year, on very small acreage, indoors, can grow what it needs to feed and clothe and provide shelter for its citizens. The citizens need only contribute according to their talents: administrative, manipulative or technical services for a small fraction of their time.

By the mid 1950s, the hoped-for agricultural revolution had not materialized, and many officials realized that nuclear energy's impact on agriculture would be less than previously thought. Yet the McKinney Panel of the Peaceful Uses of Atomic Energy suggested that nuclear energy held great promise for plant breeding, tracer research, blight and pest control, food irradiation, and crop storage. In its 1956 report to the Joint Committee on Atomic Energy, the panel recommended that continued funding, research, and technological improvements would "mean increased productivity and lower costs for individual farmers."[56] The report also admitted the obvious fact that, because cheap atomic power had not yet arrived, farming desolate wastelands was not yet possible. Essentially the same conclusions were reached at the United Nations Conference on the Peaceful Uses of Atomic Energy in Geneva in September 1958.[57]

Although there were doubts about what nuclear energy could do for agriculture, there was little question that it could be put to work in massive earthmoving and excavation projects. After all, the bomb's explosive force had been demonstrated beyond question. Certainly "peaceful" nuclear explosions could be used to blast channels, move mountains, and possibly even control the weather. Gilfillan thought that the most important use of nuclear energy would be for "shattering rock and hurling great masses of earth" by placing tiny atomic charges in "a few small drill holes." Citing the Smyth report, *Atomic Energy for Military Purposes*, Gilfillan implied that ways were already known to control atomic explosives without undue danger.[58] Perhaps the most extensive, and certainly the most enthusiastic, projection along this line was that offered by William McDermott in 1945. He suggested that "minute atomic bombs" might be used to fight fires by creating clearings and establishing firelines. Indeed, they could do the work of several thousand firefighters wielding conventional machinery, shovels, and saws. On the high seas, McDermott suggested, "Atom bombs, or well placed atomic explosive charges, offer the possibility of clearing the oceans of [icebergs]. Broken into icy slivers, they might dissolve in the warmer water and disappear. The terrific heat engendered by an atomic bomb, shown by the way sand was fused into slabs in the New Mexico experiment, indicates that heat as well as violence would aid in vaporizing the icebergs."[59] In the coming nuclear era, McDermott confidently predicted, atomic ice-blasting techniques would permit icebound ports and harbors to remain "as clear as in summer time." McDermott also suggested the use of atomic explosives to move reefs and destroy ship-menacing shoals in river channels, harbors, and sea lanes. They could also be used to create new channels and harbors. A new Panama Canal, for example, could be dug in "weeks instead of years" with small atomic explosives placed along the chosen path.[60]

Beginning in the mid 1950s, the AEC's Project Plowshare intensively studied possible uses of nuclear explosions. The Suez Crisis of 1956 prompted a meeting of scientists at Lawrence Radiation Laboratory to discuss the feasibility of using Plowshare "technology" to excavate a sea-level canal across the Sinai Peninsula. In 1958, delegates from at least a dozen nations meeting in Geneva at the Second International Conference on the Peaceful Uses of Atomic Energy discussed similar plans for large-scale earthmoving, for blasting shale to release oil, and for other purposes. Between 1957 and 1962, the AEC and the Joint Committee on Atomic Energy discussed and supported Project Chariot (a coastal harbor in Alaska), Project Carryall (a mountain pass for a rail line in California), the Tennessee-Tombigbee Project (a river canal between Alabama and Mississippi), the Cape Keranden Project (a harbor in northwestern Australia), and finally the great dream of the "Panatomic Canal" (a wider, more modern canal across the Isthmus of Panama). The Panatomic Canal project received government support from 1957 to 1970. However, after several million dollars had been spent on a study and on the exploration of alternative routes, and after discussions had been initiated with Latin American officials, the plans were dropped because of unresolved questions concerning technical feasibility, the effects of the radiation, and the overall environmental impact, and because of the opposition of the Indian tribes who lived along the right-of-way.[61]

Besides envisioning a future in which atomic energy would generate free electricity, power vehicles, and refashion the environment, the prophets believed that the application of nuclear science would help unlock the very secrets of life, including the causes of disease and of aging. In a 1946 article ("Is Atomic Energy the Key to Our Dreams?"), William Laurence wrote: "The splitting of the atom can lead to such priceless boons as the conquering of disease, the postponement of old age and the prolongation of life." These goals would be achieved through the use of so-called tagged atoms, essentially tracers that could be followed through the human body. Laurence explained:

> The "tagged atoms" of the basic elements of life that we can now create by atomic energy will enable us to trace what the animal or plant body does with its food at every stage of digestion and incorporation into the living body. They can then be used in harmless amounts in all living things, including man. By studying first the metabolism in the normal, and then comparing it with that in the various diseases, biology, medicine, physiology and biochemistry, working together, would learn for the first time what deviations take place in the sick body, and could take intelligent measures to prevent and correct these deviations.[62]

In language recalling the spiels of the patent-medicine hucksters of an earlier time, Laurence predicted that these techniques would yield cures for such dreaded diseases as cancer, arteriosclerosis, arthritis, heart disease, kidney disease, and "most other ills that are taking a tremendous toll in death and suffering." Finally, he concluded, atomic medicine would bring researchers to the "threshold of elucidating the mystery of why we get old" and would open the way for "an intelligent approach to means for postponing the process."

Laurence was not alone in viewing nuclear energy as a modern-day fountain of youth. In 1946 a *Science Illustrated* writer opined that the study of old age would be aided by radiation, and that ways would be found to slow the aging process. The writer also foresaw a reduction in neonatal mortality through the use of tracer techniques to diagnose irregularities in organisms too minute for ordinary examination. Laurence suggested that malignant growths might be stopped or retarded by depositing radioactive compounds in ailing tissue or by applying atomic radiation more powerful than x rays;[63] this prophecy was actually fulfilled.

Though occasionally the nuclear enthusiasts were right about the atomic future, more often their predictions were overly optimistic and even absurd. Many recognized this, especially with hindsight. From the vantage point of 1973, AEC Chairman James R. Schlesinger reflected back on that era of prophetic excess when many people believed "the atom could do everything except slice bread" and "there were some who thought it could slice bread."[64] But to focus on the prophets' accuracy misses a larger point. Lurking behind almost all discussions of the nuclear future lay the belief that technology based on the splitting of the atom would bring about a veritable utopia. As an expression of this ideology, the words of William McDermott in *Popular Mechanics* in 1949 are worth quoting at length:

> Unlimited power will mean the production of ample food, clothing, housing, and other necessities as well as myriad luxuries, for everyone. Poverty and famine, slums and malnutrition will disappear from the face of the earth. Disease will be attacked with fresh vigor because there is to be an ample margin of time for the cultivation of health and superb care of the body. Abundant time for study and research will make a game of mental development, and intelligence and education will reach a new high. Wars will fade out as amplitude of production of all things necessary to adequate and enjoyable living anywhere and everywhere will remove the fundamental economic and material rivalries that provoke war. Just as America, with the abundance of resources and with high standards of living, is a peaceful nation, so the spread of wealth over the poorer sections of the earth, where wars breed, will destroy the swamps where such evils generate. A new and excellent culture, the like of which the world has never before glimpsed, is likely to come into being.[65]

Clearly McDermott was interested in much more than simply the medical, transportation, or agricultural applications of nuclear energy. It was the social and cultural effects of the atom that engaged him, as they did other nuclear advocates. Their hopes in this area were probably even more misguided than their predictions regarding the technical side of the atomic future.

Technology, after all, does not ensure social progress. Sometimes new technologies can facilitate and increase an material abundance, social harmony, personal leisure, or individual freedom. But such goals cannot be achieved by simply speeding the pace of technological advance. Ideological and political obstacles, not technical ones, have blocked the path toward a better world. In this context, the dreams of an idyllic future transformed by nuclear power that captured the popular imagination from 1940 to 1970 can only be said to have raised false hopes and, even worse, perhaps diverted energies that might have gone into worthwhile efforts for social change.

Notes

1. Frederick Soddy, *The Interpretation of Radium and the Structure of the Atom* (London: John Murray, 1920), p. 172. First published in 1909.

2. Harold Stassen, "Atoms for Peace," *Ladies' Home Journal* 72 (August 1955), p. 48.

3. See William E. Akin, *Technocracy and the American Dream: The Technocratic Movement, 1900–1941* (Berkeley: University of California Press, 1977).

4. See Steven L. Del Sesto, *Science, Politics, and Controversy: Civilian Nuclear Power in the United States, 1946–1974* (Boulder: Westview, 1979).

5. R. M. Langer, "The Miracle of U-235: What Life May be Like in the Uranium Age," *Popular Mechanics* 75 (January 1941), pp. 1–5.

6. See John J. O'Neill, "Enter Atomic Power," *Harper's Magazine* 181 (June 1940), pp. 7 ff.; Harry M. Davis, "We Enter a New Era," *New York Times Magazine*, August 12, 1945, p. 43.

7. "Cheap Atomic Power in '60," *Science News Letter* 51 (February 1, 1947), p. 66. See also "What Can the Atom do for You?" *Science Illustrated* 1 (June 1946), pp. 22–27.

8. William F. McDermott, "Bringing the Atom Down to Earth," *Popular Mechanics* 84 (November 1945), pp. 1–6 ff.

9. See Public Law 585, "The Atomic Energy Act of 1946," *United States Statutes at Large* 60, part 1 (Washington, D.C.: U.S. Government Printing Office, 1947), p. 757.

10. Testimony of Robert Hutchins, hearings before Special Committee on Atomic Energy, in *A Bill for the Development and Control of Atomic Energy*, 79th Congress, 2nd session, 1946 (Washington, D.C.: U.S. Government Printing Office, 1947), p. 102.

11. William F. Ogburn, "Sociology and the Atom," *American Journal of Sociology* 51 (January 1946), pp. 267–275; Feliks Gross, "On the Peacetime Uses of Atomic Energy," *American Sociological Review* 16 (February 1951), pp. 100–102.

12. Charles E. Merriam, "On the Agenda of Physics and Politics," *American Journal of Sociology* 53 (November 1947), pp. 167–173.

13. Harry M. Fisher, "Big Things Ahead," *American Magazine* 157 (April 1954), p. 127.

14. Ibid.

15. Some of this material is based on my article "What Ever Happened to Nuclear Cars?," *Technology Review* 85 (January 1982), pp. 10 ff.

16. Louis Cassels, "Atomic Engines—When and How," *Harper's Magazine* 200 (June 1950), p. 50.

17. O'Neill, "Enter Atomic Power," pp. 7 ff.

18. S. C. Gilfillan, "The Atomic Bombshell," *Survey Graphic* 34 (September 1945), p. 358.

19. R. M. Langer, "The Miracle of U-235: What Life May be Like in the Uranium Age," *Popular Mechanics* 75 (January 1941), p. 4.

14. Ibid., p. 5.

21. R. M. Langer, "Fast New World," *Collier's* 106 (July 6, 1940), p. 54.

22. Ibid.

23. Langer, "The Miracle of U-235," p. 4. A. M. Low, president of the British Institute of Technology and designer of the first radio-controlled plane, believed that atomic energy could be utilized to drive cars in a somewhat different way. He said electronic cars could be driven by transformers on electrical highways if ways could be found of tracing what he called "the exploding specks of energy" that would be released. See A. M. Low, "What's Next with the Atom?" *Popular Science Monthly* 147 (October 1945), p. 66.

24. Langer, "The Miracle of U-235," p. 150A.

25. E. V. Murphee, "Power from Atoms: How Soon?," *Popular Mechanics* 86 (October 1946), p. 93.

26. "Future Atomic Advances," *Science News Letter* 53 (May 22, 1948), p. 326.

27. Out of the Magic of A-Power: Things to Come," *Newsweek*, September 19, 1960, p. 69. See also "Extraordinary Atomic Plane: The Fight for the Ultimate Weapon," *Newsweek*, June 4, 1956, p. 58.

28. Harry S. Baer, Jr., "Nuclear Power for Aircraft," *Flying* 38 (March 1946), pp. 21 ff.

29. Langer, "Fast New World," p. 54.

30. "Atomic Airplane," *Newsweek*, January 28, 1948, p. 54. See also "Atom-Driven Planes," *Time*, July 5, 1948, p. 44.

31. "Atomic Power for Future Super-Airliners," *Science News Letter* 49 (April 20, 1946), p. 248.

32. See also O'Neill, "Enter Atomic Power."

33. Harold Wolff, "A World Worth Waiting For," *Coronet* 25 (November 1948), p. 34. Wolff did qualify the statement later in the article by noting some of the difficult problems still to be solved.

34. Alvin Weinberg, "Peacetime Uses of Nuclear Power," *New Republic* 114 (January 25, 1946), p. 276.

35. Gilfillan, "The Atomic Bombshell," p. 358.

36. William L. Laurence, "Is Atomic Energy the Key to Our Dreams?," *Saturday Evening Post* 218 (March–April 1946), p. 41.

37. McDermott, "Bringing the Atom Down to Earth," p. 164.

38. Harry M. Davis, "We Enter a New Era," *New York Times Magazine*, August 12, 1945, p. 43.

39. Langer, "The Miracle of U-235," pp. 2–3.

40. Luis Alvarez, "Atomic Energy for Space Ships?" *Science Digest* 21 (May 1947), pp. 84–85. Emphasis in original.

41. Ibid.

42. Leonard I. Katzin, "Industrial Uses of Atomic Energy," *Scientific American* 174 (February 1946), p. 75. Other rebuttals can be found in Murphree, "Power From Atoms" and W. L. Davidson, Jr., "We *Can* Harness the Atom," *Popular Science* 147 (December 1945), pp. 65–69.

43. W. Henry Lambright, *Shooting Down the Nuclear Plane* (Indianapolis: Bobbs-Merrill, 1967), pp. 23–24.

44. See "The Outer Space Development Amendment of 1958," in Hearings before the Subcommittee of the Joint Committee on Atomic Energy, *Outer Space Propulsion by Nuclear Energy*, 85th Congress, 2nd session, 1958 (Washington, D. C.: U.S. Government Printing Office, 1958), pp. 199–200.

45. Hearings before the Subcommittee on Research, Development, and Radiation of the Joint Committee on Atomic Energy, *Nuclear Energy for Space Propulsion and Auxiliary Power*, 87th Congress, 1st session, 1961 (Washington, D.C.: U.S. Government Printing Office, 1961), p. 184. Senator Anderson said the program should contain the following elements:

(1) A test flight of a nuclear rocket prototype, including operation of facilities for the generation of electricity in outer space by nuclear energy should be completed by the end of 1964. In this mission the great payload advantage of nuclear power should be demonstrated. (2) A manned expedition to the moon utilizing nuclear rockets and nuclear electric power sources should be undertaken as soon as possible before the end of the 1960s. (3) Maximum development possible in nuclear rockets and nuclear electric power sources (in excess of one megawatt) is to be undertaken so that by the end of the 1960s more extensive explorations of our solar system can be programmed.

46. *Nucleonics Magazine*, April 1961. Cited in: Subcommittee on Research, Development, and Radiation, *Nuclear Energy for Space Propulsion and Auxiliary Power*, p. 184. For additional materials on ideas of nuclear flight, see R. W. Bussard and R. D. Delauer, *Fundamentals of Nuclear Flight* (New York: McGraw-Hill, 1965); Kenneth F. Gantz, ed., *Nuclear Flight* (New York: Duell, Sloan, and Pearce, 1960).

47. "Atomic Research May End the World's Hunger," *Christian Century* 65 (July 28, 1948), pp. 749–750.

48. Cassels, "Atomic Engines," pp. 50–53.

49. See, e.g., "What's Ahead for the Atom," *Science Digest* 23 (May 1948), pp. 88–91.

50. Laurence, "Is Atomic Energy the Key to Our Dreams?," p. 41. See also William L. Laurence, "Paradise or Doomsday?," *Women's Home Companion* 75 (March 1948), pp. 32 ff.

51. Wolff, "A World Worth Waiting For," p. 34.

52. "Atoms May Revolutionize World Social Order," *Science News Letter* 49 (May 4, 1946), p. 277.

53. Gilfillan, "The Atomic Bombshell," p. 358.

54. "What's Ahead for the Atom," p. 88.

55. Langer, "Fasts New World," pp. 54, 19.

56. Joint Committee on Atomic Energy, 84th Congress, 2nd session, 1956, *Peaceful Uses of Atomic Energy: Report of the Panel on the Impact of the Peaceful Uses of Atomic Energy* (Washington, D.C.: U.S. Government Printing Office, 1956), p. 63.

57. See *Proceedings of the Second United Nations International Conference on the Peaceful Uses of Atomic Energy*, volume 27, "Isotopes in Agriculture" (Geneva: United Nations, 1959).

58. Gilfillan, "The Atomic Bombshell," pp. 357–358.

59. McDermott, "Bringing the Atom Down to Earth," p. 6.

60. Ibid.

61. Richard S. Lewis, *The Nuclear Power Rebellion: Citizens Versus the Atomic-Industrial Establishment* (New York: Viking, 1972), pp. 172–198. See also E. A. Martell, "Plowing a Nuclear Furrow," *Environment* 11 (April 1969), pp. 3–12 ff.

62. Laurence, "Is Atomic Energy the Key to Our Dreams?," p. 41.

63. Ibid., p. 23.

64. Hearings before the Joint Committee on Atomic Energy, *The Status of Nuclear Reactor Safety*, 93rd Congress, 1st session, 1973 (Washington, D.C.: U.S. Government Printing Office, 1974), p. 8.

65. McDermott, "Bringing the Atom Down to Earth," pp. 6 ff.

4

Plastic, Material of a Thousand Uses

Jeffrey L. Meikle

So much of our contemporary environment is molded, woven, fabricated, or constructed of plastics that normally we hardly notice their existence as a distinct class of materials. Few of us are aware that the "Plastics Age" arrived in 1979,[1] when the annual volume of plastics produced exceeded that of steel. We just don't care. Satellite TV antennas and home computers monopolize our interest in domestic applications of technology. When we do think about plastics, we are apt to react negatively. We may worry about discarded polyethylene bottles washing up on our favorite beaches, or about fragments of styrofoam clogging the Sargasso Sea. We may fear the onset of cancer induced by a Teflon coating that gradually disappeared, presumably into our food, from the surface of a long-discarded frying pan. Some of us affect a preference for "natural" materials such as wood, leather, cotton, and wool. But all of us are surrounded by plastics. We use plastic objects countless times every day, even those of us who laughed at the lines about plastics in the opening scene of *The Graduate* in 1967 and who learned to use the adjective *plastic* as a synonym for *false*, *phony*, or *superficial*.[2]

Earlier in this century, however, when the plastics industry was in its infancy, plastics enjoyed a highly favorable reputation. From about 1920 to 1950, the idea of a Plastic Age conjured up utopian visions of an infinite array of synthetic materials and products to be provided inexpensively, literally out of thin air, by the wizards and miracle workers of industrial chemistry. Ironically, and perhaps inevitably, it was only after plastics had become ubiquitous that this utopian dream turned sour. Public interest in plastics as a miracle material declined as products made of plastics became more and more commonplace in every area of daily life. But as long as the development of synthetic materials seemed part of the leading edge of scientific and technological advancement, and as long as the industry's spokesmen could promise ever more extravagant applications for new plastics, Americans continued to regard them as the stuff from which a decidedly materialistic utopia would be molded.

One of the most complete expressions of this plastics utopianism appeared in December 1941 (on the eve of Pearl Harbor) in *Science Digest*, a magazine devoted to simplified popularizations of science and technology. Two applied chemists, V. E. Yarsley and E. G. Couzens, contributed an exercise in impressionistic prophecy entitled "The Expanding Age of Plastics,"[3] which traced the life of an imaginary "Plastic Man," a "dweller in the 'Plastic Age' that is already upon us." As an infant, he would "come into a world of color and bright shining surfaces, where childish hands find nothing to break, no sharp edges or corners to cut or graze, no crevices to harbor dirt or germs." He would enjoy walls, toys, a crib, a teething ring, bottles, spoons, and mugs of plastic, and would be "surrounded on every side by this tough, safe, clean material which human thought has created." Later, as a child, the young Plastic Man would undergo education, in a "new kind of schoolroom" with "shining unscuffable walls," at a "moulded desk, warm and smooth and clean to the touch, unsplinterable, without angles or projections." In his home, the "universal plastic environment," Plastic Man would live amid synthetic veneers, plastic bathroom fixtures and pipes, molded furniture, lamps and screens of "beautiful transparent glass-like materials," and "bouquets of flowers miraculously preserved in geometrical masses of brilliant, clear, transparent plastic." Although his automobile would be "almost wholly plastic in its externals," Yarsley and Couzens passed quickly over its virtues to describe Plastic Man's personal airplane as "the motor-car of the future," with all of its parts "mass-produced from reinforced plastic." Synthetics might universalize the power of flight, but they unfortunately could not stave off "the inevitable end." Old age would be brightened, however, by plastic teeth, glasses, playing cards, and chessmen, and Plastic Man eventually "sinks into his grave hygienically enclosed in a plastic coffin."

Forty years later, this extravagantly utopian description of Plastic Man and his environment seems almost a caricature of the world-of-tomorrow magazine piece. Yarsley and Couzens took their utopian vision seriously, however. They found in the future Plastics Age an opportunity for the human race to extend control over its own destiny, to gain freedom from chance, to eliminate the imperfections of nature, and even to put an end to wars such as the one that had engulfed Europe. The world of Plastic Man would be "brighter and cleaner" than any previously known, "a world free from moth and rust and full of color." Reversing the main fact of economic life since the beginning of human history, it would be a place of abundance rather than scarcity—"a world in which man, like a magician, makes what he wants for almost every need, out of what is beneath him and around him, coal, water and air." Perhaps most important of all, the new civilization, "built to order" by industrial chemists and technocrats, would be "the perfect expression of the spirit of scientific control, the Plastics Age."[4]

Saving civilization constituted a heavy burden for any industry to bear, and at no other time before or since did anyone expect it of the plastics industry. Even in 1941, most Americans had encountered synthetic materials in breakfast-cereal premiums, costume jewelry, radio cabinets, automobile steering wheels, Nylon stockings, and Formica counter tops. Taken separately, none of these applications and developments seemed exactly utopian in scope (with the possible exception of Nylon, which Dupont's vice-president for research had unwisely claimed to be "as strong as steel, as fine as the spider's web"[5]). And yet Americans did view the plastics as miracle materials from which they would shape the precise contours of a desired future. The story of how plastics first gained and then lost their utopian significance not only forms a chapter in the history of American utopianism; it also illuminates the cultural intersection of aesthetics, economics, and technological innovation in the twentieth century.

The first true plastics, derived from nitrocellulose, appeared in the 1860s. But no one recognized them as revolutionary new materials. In fact, their inventors and promoters conceived of them merely as substitutes for more expensive materials. When in 1865 the Englishman Alexander Parkes first described his substance Parkesine, which he had developed three years earlier, he noted that he had for twenty years been looking for a material "partaking in a large degree of the properties of ivory, tortoise-shell, horn, hard wood, india rubber, gutta percha," and so on, and intended to "replace such materials."[6] The more famous John Wesley Hyatt, an American who in 1869 discovered celluloid, made from nitrocellulose and camphor, also intended his product as an imitative substitute for other materials and not as a new material to be exploited for its own unique properties and virtues. In 1936, at the height of the plastics craze, *Fortune* magazine accorded Hyatt the almost biblical distinction of "having created a new substance under the sun, a material that was not to be found in nature and that could not be converted back again into the substances out of which it was made."[7] Hyatt, however, lacked such a vivid imagination, and concentrated on imitating the natural materials used in making billiard balls, false teeth, buttons, combs, handles for brushes and razors, and detachable shirt collars. Imitation became as exact as Hyatt and his patent licensees could make it. Imitation-ivory celluloid jewelry boxes, for example, bore the delicate swirls indicative of the growth of an elephant's tusk; celluloid collars imitated the weave and stitching of linen collars. As historian Robert Friedel has aptly phrased it, "the first important function of this plastic was to look and behave like something it was not."[8] Thus, as early as the nineteenth century we find a negative image of plastic as a cheap, imitative, second-rate substitute material—an image that the plastics industry later fought to overcome by stressing the unique virtues of its synthetic products.

At first, not even Leo Baekeland, an immigrant chemist from Belgium who produced the first truly synthetic plastic in 1907, had any inkling of his discovery's revolutionary significance. Among his many research projects, Baekeland sought a synthetic substitute for shellac, a natural resin manufactured from insect secretions that was in great demand for its electrical insulating properties. Joining the compounds phenol and formaldehyde under controlled conditions of heat and pressure, Baekeland created a synthetic resin with a unique molecular structure of its own. Always a businessman as well as a chemist, he deferred announcing his discovery to the scientific and industrial communities until he had received a patent on this new substance, which he christened Bakelite. At first Baekeland conceived of his phenolic resin primarily as a coating or fixative. It could usefully be applied, he asserted, as an electrical insulator, as a protective coating for wood "superior to any varnish and even better than the most expensive Japanese lacquer," as a means of turning "cheap, porous soft wood" into "a very hard wood, as hard as mahogany or ebony," and as a rustproof coating for metals. Only as an afterthought did Baekeland admit that his new material might find application in consumer goods, as a replacement for amber or flammable celluloid in pipe stems or for ivory in "knobs, buttons, knife handles." He concluded, however, that "[Bakelite's] use for such fancy articles has not much appealed to my efforts as long as there are so many more important applications for engineering purposes." Any applications of Bakelite to consumer goods, he assumed, would be relatively trivial and would, like the use of celluloid, represent substitution and imitation.[9]

American manufacturers soon wondered how they had ever done without Bakelite. Whether used as a varnish or shellac, as an impregnating binder and hardener of laminated wood and paper products, or as a molding compound for small objects (with bits of wood, cloth, or asbestos as a filler), Bakelite quickly found innumerable markets.

Its strength, hardness, insolubility, resistance to acids, and insulating properties with regard to heat and electricity made it the material of choice, rather than a cheap substitute, for many industrial uses. When the Society of the Chemical Industry awarded Leo Baekeland its highest honor in 1916, the official presenter cited the use of Bakelite "for the most varied purposes, ranging from the manufacture of a billiard ball to that of a wireless telegraphic apparatus; from the manufacture of a self-starter for automobiles to that of transparent fountain pens," and including such other things as "switchboards for battleships, moldings for kodaks, phonograph records, casings for instruments of precision, armatures and commutators for dynamos and motors, telephone receivers, railroad signals, grinding wheels, umbrella handles, buttons, cigar holders and pipe stems, articles of ornament, etc."[10] Even into the late 1920s, the vast majority of applications of Bakelite to consumer goods occurred in the hidden electrical parts of automobiles and radios. Although the plastics industry developed in tandem with the automotive and electrical industries, it did not immediately share in their futuristic machine-age glamour. A Bakelite distributor cap, however functional and essential to progress it might have been, did not stimulate utopian aspirations. By that time, however, plastics shared in a utopian mystique that the larger chemical industry had begun to generate soon after the turn of the century.

In 1907 the readers of *Everybody's Magazine* were informed about "the new synthetic chemistry" in an article entitled "The Miracle-Workers: Modern Science in the Industrial World."[11] Focusing primarily on the fixation of nitrogen to provide artificial fertilizer, the author concluded that "the raw materials for building up" just about anything "lie everywhere about us in abundance." Scientists would never be satisfied until they could fabricate "a loaf of bread or . . . a beefsteak" from "a lump of coal, a glass of water, and a whiff of atmosphere." Numerous other articles in the popular press echoed this refrain, but the theme became much more frequent during and after World War I—which was, according to one observer, "a war of chemists against chemists."[12] Over the years, these articles celebrating the wonders of industrial chemistry paid increasing attention to the plastics, including celluloid, Bakelite, and casein (a derivative of skim milk used primarily in making buttons).

Edwin E. Slosson, a prolific popularizer of chemistry, repeated his usual points in an article on "Chemistry in Everyday Life" published in *The Mentor* in 1922.[13] The human race, according to Slosson, stood at the beginning of an era of unprecedented material abundance made possible by coal—dirty, black coal, one of the most aesthetically unappealing substances known to man. This limitless "scrap heap of the vegetable kingdom" provided the chemist with "all sorts of useful materials" because coal contained in condensed form "the quintessence of the forests of untold millenniums." From sticky, evil-smelling coal tar, scientists synthesized not only dyes in "all the colors of Joseph's coat" and life-saving medicines but also phenol, one of the two major ingredients of Bakelite. Thanks to Bakelite and to celluloid (characterized by Slosson as a "chameleon material"), imitations of amber, ebony, onyx, and alabaster were "now within the reach of everyone." According to Slosson, these imitation materials actually surpassed the expensive originals in both appearance and usefulness. The synthetic chemist thus acted as an "agent of applied democracy" by making apparent luxuries available as "the common property of the masses."

One could hardly imagine a more utopian celebration of industrial chemistry and the new plastics, but John Kimberly Mumford managed to evoke a tone of millennial expectancy in a little book devoted entirely to Bakelite, published in 1924.[14] Mumford was a journalist who focused on popularizing industrial topics. He opened his story with the creation of life itself, when nature first began storing up the "waste heaps" of dead organic matter from which research chemists were later to derive "colossal assets." Bakelite appeared as "a wonder-stuff, the elements of which were prepared in the morning of the world, then laid away till civilization wanted it badly enough to hunt out its parts, find a way to put them together and set them to work." Mumford marveled over Bakelite's "Protean adaptability to many things" and the ease with which it could be molded, but he found even more miraculous the chemical reaction by which it set or hardened, after which it would "continue to be 'Bakelite' till kingdom come." Out of the very stuff of death, Mumford seemed to be suggesting, Leo Baekeland had created an indestructible material imbued with immortality. And the chemists who followed him would indeed "make a new world"; they would "create new substances, and create them out of anything." Mumford provided a long list of applications of Bakelite, many of them for external use in consumer products. He concentrated most, however, on Bakelite's power as an electrical insulator. This "super-resin," as he referred to it, marked the "answer of the Chemist to the call for a new rein on Electricity, the tricky, but titanic force that pulls the apple-cart of the human race." Bakelite certainly took on utopian significance in Mumford's view, but it remained a material used primarily behind the scenes, invisible to the average person yet of supreme importance to the functioning of industry.

With increasing frequency in the 1920s and the 1930s, popular magazines and books celebrated synthetic chemistry—and, by association, the plastics—as a sort of utopian magic, capable of creating a world of transcendent beauty and perfection from the most common of earth's elements. Even *Fortune*, the most intellectual of the nation's business journals, entitled its 1936 review of the plastics industry with a biblically resonant phrase: "What Man Has Joined Together. . . ."[15] A more practical journal, *Business Week*, attributed the hoopla surrounding plastics to public ignorance, a result of "the difficulty of translating into common terms the mysterious ways in which chemical processes move, their wonders to perform."[16] Lest anyone take seriously this use of biblical language, *Business Week*'s anonymous correspondent savagely attacked "popular stories of the 'modern miracle' type, illustrated by figures symbolizing commercial research holding test tubes (usually by the wrong end)." This satirical comment merely highlighted the degree to which the public did believe in modern miracles, ones which could be directed and controlled by human agency.

The popular view of the plastics as the miracle materials of a futuristic machine age received perhaps its most extravagant expression—with a highly significant twist in emphasis—in *Form and Re-form*, a book published by industrial designer Paul T. Frankl in 1930.[17] "Base materials," Frankl wrote, "are transmuted into marvels of beauty" by industrial chemistry, which "today rivals alchemy" in its achievements. Plastics such as Bakelite "spoke" to modern man in "the vernacular of the twentieth century"—"the language of invention, of synthesis." Unlike the pioneers Parkes, Hyatt, and Baekeland, who thought of their discoveries primarily in terms of imitation of more expensive natural materials, Frankl asserted "the autonomy of new media." It was necessary to recognize the revolutionary nature of the plastics and to "create the grammar of these new materials." To put it bluntly, plastics could express the unique beauty of a new age. As an industrial designer, Frankl was well aware that plastics were becoming more visible in design and architecture—areas where style, taste, and aesthetics received paramount attention. As plastics shed their almost exclusively industrial image and moved into the realm of consumer products, they took with them the aura of utopian magic with which the public, helped along by enthusiastic journalists, had imbued them.

Many factors contributed to the rise of plastics as consumer-goods materials in the late 1920s and into the Depression years. Among the most significant were increasing competition among plastics suppliers, the development of new types of plastics, and, perhaps most important, the rise of industrial design as a profession during the Depression.

Competition first became a factor in 1927 when Leo Baekeland's patent on phenolic resin expired. As other manufacturers rushed to open phenolic plants, the artificially controlled price of Bakelite went down, and it became economically feasible to use the material for relatively frivolous or inexpensive consumer products, in addition to industrial applications for which cost was really no object.[18] Competition also stimulated the introduction of new phenolic plastics in a rainbow of colors—plastics which were more suitable than Bakelite for consumer goods intended to appeal to the American woman. The Bakelite Corporation had traditionally supplied its product in black in order to disguise impurities in the fillers used to give it strength in molded form. In 1927, taking advantage of the patent expiration, the American Catalin Corporation announced the availability of "Catalin, an insoluble, infusible cast phenolic resin of gem-like beauty and an unlimited color range, which in form of rods, tubes, sheets, or shapes can be machined on ordinary shop equipment as easily as wood or brass."[19] Used only for such things as small toys, costume jewelry, chessmen, and decorative panels, Catalin did not require a filler for strength and could be supplied in virtually any solid, mottled, translucent, or transparent color. Not to be outdone, the Bakelite Corporation immediately offered its own line of cast resins, which supposedly made possible "the reproduction of precious stones."[20]

The end of the 1920s also witnessed the development of entirely new plastics to compete with the molded and cast phenolics. Cellulose acetate, developed by the Celluloid Corporation and marketed under the trade name Lumarith, was a colorful, nonflammable celluloid substitute often used in lighting fixtures and lampshades.[21] Vinyl, introduced in 1928 under the name Vinylite, was immediately used for phonograph records, dentures, and the lining of beer cans; however, it gained its most spectacular exposure at the Chicago Century of Progress Exposition of 1933, which boasted a "house of the future" in which the interior walls, the floors, the furniture, and many of the fixtures were made of vinyl.[22] The most important of the new plastics were the urea formaldehyde resins, developed specifically as molding compounds that, unlike Bakelite, could simultaneously furnish both strength and an infinite range of possible colors, including white. The first of these urea formaldehyde plastics was made available in the United States in 1929 by the American Cyanamid Company under a license granted by a British firm. The material's unlikely trade name—Beetle—was soon attached to the first set of plastic dishes offered for sale in America—Beetleware.[23] Almost simultaneously, the Toledo Scale Company funded research at the Mellon Institute in Pittsburgh aimed at developing a urea plastic strong enough and light enough in color to be used for the housing of a grocery-counter scale. The resulting compound, Plaskon, went on the market in 1931, but production of the scale—which required a 45-ton molding press, the largest then in existence—did not begin until 1935.[24] Rounding out the roster of plastics available before World War II were moisture-proof cellophane, introduced for packaging by Dupont in 1927; the clear, glasslike acrylics, marketed as Plexiglas by Röhm & Haas and as Lucite by Dupont (these were the first plastics to be derived from petroleum); and of course Nylon, the "sheer magic" of which was first tested by the young women who guided visitors through Dupont's pavilion at the 1939–40 New York World's Fair.[25] But the popularity of plastics did not derive entirely from the availability of new varieties with diverse characteristics. Strong economic motives dictated the use of plastics in consumer products and at the same time ensured that product designers would employ plastics in futuristic machine-age styles that would enhance their utopian aura.

In a very real sense, the plastics industry and the new profession of industrial design developed together during the Depression years. Industrial designers—among them Walter Dorwin Teague, Henry Dreyfuss, Raymond Loewy, Norman Bel Geddes, and Harold Van Doren—spent the decade redesigning America's consumer products to make them more attractive to potential purchasers. Businessmen hoped that industrial design would overcome the so-called underconsumption problem, which they considered the root of all the nation's economic woes. Innumerable products, from toasters to refrigerators, underwent face-lifting operations in an attempt to engineer economic recovery. If, as industrial designers liked to say, their profession was a "depression baby,"[26] then the plastics industry was, in *Fortune*'s words, a "child of the depression."[27]

Despite the hyperbole, there was some truth to a trade journal's assertion that "A Plastic a Day Keeps Depression Away."[28] Industrial designer Peter Müller-Munk later correctly claimed that during the 1930s "plastics became almost the hallmark of 'modern design'—as the mysterious and attractive solution for almost any application requiring 'eye appeal.' "[29] There were several good reasons for this. In the first place, in order to make their products more affordable by lowering prices, manufacturers had to find ways of reducing production costs. As plastics became cheaper, they became increasingly attractive as substitutes for such traditional materials as metals, ceramics, and wood. In addition, products made from molded plastics often did not require expensive, labor-intensive assembling and finishing operations.

A plastic radio case, for example, popped out of its mold in a single piece, already brightly colored by a dye that had been incorporated into the molding compound. Manufacturers hesitated only because of a lingering feeling that plastics represented imitative substitutes which the public would reject as inferior to more natural or traditional materials. The plastics industry itself solved the problem by mounting an aggressive campaign intended to prove that synthetic materials possessed exactly the aesthetic qualities required for design in a utopian machine age. Franklin E. Brill of General Plastics summed up the focus of the campaign when he urged his colleagues to avoid "such time-worn motifs as colonial silhouettes and Scottish terriers" in order to "make decoration symbolic of our modern age, using simple machine-cut forms to get that verve and dash which is so expressive of contemporary life."[30]

The Bakelite Corporation, already under competitive attack, led the push to convince manufacturers to beautify their products by using plastics. In 1932 the firm held a symposium to acquaint industrial designers with the technical advantages and limitations of plastics as materials and also with their stylistic potential. Many businessmen already held in awe the figure of the industrial designer, considering him, as one designer later recalled with a touch of bitterness, "a wizard of gloss, the man with the airbrush who could take the manufacturer's widget, streamline its housing, add a bit of trim, and move it from twentieth to first place in its field."[31] If industrial designers could be won over to plastics, it seemed, then the battle was won. They needed little convincing. Over the next two years, trade journals such as *Modern Plastics* and *Sales Management* featured numerous advertisements focusing on individual designers and their Bakelite products. Each ad spotlighted a single product, each contained a small photograph and capsule biography touting the designer as celebrity, and each quoted the great man himself on the virtues of sleek, modern, machine-age design. The various applications of Bakelite included small knobs and handles, personal accessories such as barometers and telephone indexes, irons, washing machines, mimeograph machines, and soft-drink dispensers. Through all these ads ran the message that Bakelite—billed as "the material of a thousand uses"—would both revitalize industry and help remake the environment in the image of a utopian world of tomorrow, beyond the chaos of the Depression.

Other plastics suppliers echoed Bakelite's ambitious campaign. Even more important, by the end of 1934 *Modern Plastics*, the industry's major trade journal, had turned its editorial focus away from technical matters and toward industrial design. With a slick new layout provided by industrial designer Joseph Sinel, the magazine featured interviews with industrial designers, glossy photographs of products redesigned in plastics, product success stories, and tips on how to use plastics to their best stylistic and aesthetic advantages. No longer intended for an audience of materials suppliers, molders, and fabricators, *Modern Plastics* had become a highly effective means of convincing businessmen in other industries that they might safely and profitably switch to plastics as the raw materials of their products.

The relationship between the plastics industry and industrial design was indeed symbiotic. As *Business Week* rather awkwardly phrased it in 1935, "modernistic trends have greatly boosted the use of plastics in building, furniture and decoration, and contrariwise, plastics by their beauty have boosted modernism."[32] As it turned out, the most economical methods of making molds for plastic products lent themselves quite nicely to the major styles of utopian or machine-age styling that were current during the Depression. For example, Harold Van Doren's Air-King radio of 1933, one of the first radios to have a plastic case, exhibited the characteristic zigzag setbacks and geometric ornamentation of the style we now call Art Deco. As Van Doren himself pointed out, geometric designs—"steps, ribs, circles, flutes, etc."—could be cut into a mold by machine tools, thus eliminating the prohibitive expense of hand labor and craftsmanship.[33] To design in any more intricate fashion—imitating the shapes of the traditional woodworker, for example—would have made the production of molds too expensive. Thus, of necessity, plastics became identified in the consumer's mind with modern life and machine-age styling.

Even more appropriate to the technology of molding plastics was streamlining, the new aerodynamically derived style that began to replace the jagged Art Deco style during the mid 1930s. Low, horizontal, sculptural, flowing, and evocative of speed, streamlined design reflected the desire of Americans for smooth, frictionless flight into a utopian future in which rounded vehicles, machines, and architecture would provide a protective, uncomplicated, harmonious environment closed off from the sorts of social and economic dislocations that marked the Depression. As streamlining spread from locomotives and automobiles (where it supposedly reduced the retarding friction of wind resistance) to stationary products such as pencil sharpeners and meat slicers, it came in for a good bit of criticism from design purists. But the public loved streamlining, and defenders of the style could argue that in the area of molded plastics it made good sense. Even as late as 1946, nearly every authority on plastic product design emphasized the virtues of streamlining. A rounded or streamlined mold could be cut and

Figure 1
This 1940 Crosley table radio, made of Bakelite, displays the bulbousness that characterized the mature phase of streamlined styling. *Modern Plastics* 17 (February 1940), p. 32.

polished by machine, but the angles of a mold with sharp edges and corners required expensive hand finishing. A rounded mold also permitted smooth, even flow of the plastic material to every area and surface. Copying the wind-tunnel tests of aerodynamic engineers, plastics engineers even used special dyes to trace the flow of plastics through variously shaped molds.[34] Such experiments led to the conclusion that "streamlined flow should be designed into both the inside and outside" of a plastic part in order to avoid gas pockets, uneven flow, and other flaws that could weaken the finished product.[35] Once a plastic radio or electric razor was in the hands of the purchaser, rounded edges and corners would provide greater protection from breakage than thinner, sharper angles and corners. And, in the final analysis, rounded contours brought out the reflective beauty of glossy plastic as a material. Since plastic lent itself to flowing shapes, it should be used sculpturally. Plastics and streamlining thus went together like hand and glove. One journalist even claimed that the requirements of molding technology had inspired streamlining as a design style.[36] Such a statement was farfetched, but there can be little doubt that, as the 1930s came to a close, the "miracle materials" were all the more closely linked, through industrial design and styling, to the concept of an approaching technological utopia.[37]

Faced with a veritable explosion of plastic materials and futuristic applications, the average American of the 1930s could hardly keep from investing plastics with utopian potential. Popular journalism presumably reflected typical attitudes about the new materials. In 1932, the worst year of the decade, the *Literary Digest* ran an article on the "Approach of the 'Plastic Age,' "[38] and the *Review of Reviews* tagged a summary of plastics developments with the hopeful title "Synthetic Age . . . Era of Make-Believe."[39] A few years later, *Arts and Decoration* unimaginatively described an approaching "Plastic Age,"[40] and *Popular Science Monthly* returned to the rhetoric of a decade earlier in revealing how "New Feats of Chemical Wizards Remake the World We Live In."[41] At the end of the decade, *National Geographic* characterized a fashion model "clad from head to foot in artificial materials" as "a startling symbol of this new artificial world risen so fast since the World War"[42]; *Popular Mechanics* made the hackneyed observation that "with plastics invading one field after another, we seem to be emerging into what might be called 'The Plastics Age.' "[43] *PM* redeemed itself, however, by boldly claiming that "the American of tomorrow . . . clothed in plastics from head to foot . . . will live in a plastics house, drive a plastics auto and fly in a plastics airplane."[44]

Along these lines, plastics utopianism received an eccentric but crucial boost from Henry Ford in November 1940. Ford, who was then 77 years old, picked up an ax and swung it as hard as he could into the trunk lid of a custom-built 1941 automobile as reporters took notes and photographers recorded the event for posterity.[45] Rather than crumple and lose its paint, the lid rebounded and looked as good as new because it was made of a tough plastic with its color embedded. Constructed of vegetable fibers (hemp, flax, and ramie) compressed in a phenolic resin, the panel partially fulfilled Ford's dream that "some day it would be possible to grow most of an automobile."[46] Ford desired to relieve the economic problems of American farmers by creating an industrial market for agricultural products, and since 1930 his company had been developing a casein-type plastic from soybeans. Even though his soybean plastic found only minor applications, Ford unveiled a car whose body was made entirely of fiber-phenolic panels in August 1941 and predicted that within a year or two plastic cars would be rolling off the assembly line.[47] Had the war not intervened, the Ford Motor Company would have discovered the impossibility of fastening plastic panels to a tube-steel frame quickly enough to permit true mass production. But Ford's plastic car was never put to the test. As the United States approached entry into the war, new developments in plastics occurred under extreme secrecy,

Figure 2
Henry Ford swinging an ax into the plastic trunk panel of an experimental Ford automobile late in 1940. Ford Archives/Henry Ford Museum, Dearborn, Michigan.

and Americans were left with the notion, expressed by a *Time* correspondent, that "the technological novelty known as plastics" had "graduated from its celluloid-and-Beetleware phase into an instrument of industrial revolution."[48] Almost everything, it seemed, would be made of plastics once the war was over.

This optimistic attitude about postwar uses of plastics was evident in a little book by B. H. Weil and Victor J. Anhorn, *Plastic Horizons*, published in 1944 as part of a series on "Science for War and Peace." The authors cited a typical wartime advertisement from *Modern Plastics* asserting: "When the Minute Man returns to his Plow—it will have *Plastic* handles!"[49] The plastics industry expanded tremendously during the war, primarily to provide essential military equipment but also to replace traditional metals that were no longer available for the most basic consumer goods. Every G.I. encountered plastics in his phenolic helmet liner, vinyl raincoat, ethyl cellulose canteen, urea buttons, cellulose acetate bayonet scabbard and gas mask parts, and melamine

Figure 3
Industrial designer Carl W. Sundberg tantalized Americans with this vision of an all-plastic auto trailer/camper/motorboat shell. *Modern Plastics* 22 (May 1945), p. 106.

dishes.[50] The general public, however, was more interested in Plexiglas airplane canopies and noses, which seemed to foreshadow bubble-domed postwar automobiles of teardrop shape, and in laminated-plastic gliders and light reconnaissance planes, which held the promise of "the family car of the air" or "the Ford of the skyways."[51]

During the war, journalists and advertisers built up so many utopian predictions about plastics in the postwar world that the industry could never have come close to fulfilling them. Magazine advertisements promised the folks on the home front everything from plastic "Supercars" to a plastic "Supermix" food mixer—hot off the industrial designers' drawing boards and "just waiting for America to win the right to enjoy them."[52] "They won't be miracles tomorrow," asserted the headline of another ad,[53] and *Newsweek* announced: "Test-Tube Marvels of Wartime Promise a New Era in Plastics."[54] Combining the older utopian vision of the white-coated laboratory chemist and the newer vision of a postwar plastic cornucopia, *Newsweek* marveled over the fact that the "molecule engineers" could "almost draw blueprints of the kind of new molecules that they need for a given purpose." But plastics utopianism promised more than mere molecules—even synthetic ones—could deliver.

Plastics did make up an ever-greater percentage of the materials used in consumer goods during the postwar years, but their reputation took a catastrophic turn for the worse. There were several reasons for the demise of the utopian image of plastics. Most obvious, cars and houses were not molded of plastics, and the family airplane, of whatever material, never became a reality. More significant were some less obvious factors. Plastics had served during the war as substitutes for other materials in consumer goods—often in applications for which they were not truly appropriate. During the war, people sometimes tended to consider plastics, according to one account, as "ersatz—something to be worried along with until more common materials are once more obtainable."[55] Once the war had ended, people wanted "genuine" materials. A shop on Fifth Avenue, for example, reported that its customers preferred to "wait for hard-to-get leather luggage rather than accept similar long-wearing, good-looking plastic styles."[56] To make matters worse, many of the plastic products sold at war's end were made from poor-quality scrap by manufacturers who either did not know or did not care that often they failed to match a particular type of plastic with its appropriate uses. Consumers complained bitterly about combs that dissolved in hair oil, dishes that softened and lost their shapes in hot water, buttons that became greasy blobs during dry cleaning, and hairbrushes that broke in two after being dropped a few feet.[57] The industry made a concerted effort to discipline unscrupulous manufacturers, to educate ignorant ones on the properties of various plastics, and to eliminate an undercurrent of opinion that plastics were inherently shoddy. Although the quality of plastic products improved continually, changing preferences in style and design undermined these efforts to salvage the popular reputation of plastics.

Frankly synthetic materials, so popular during the 1930s, no longer appealed to Americans. As they attempted to come to terms with the postwar world, they turned from the machine-age styles of Art Deco and streamlining to seemingly more traditional, pseudo-historical styles reflecting the American past. The stark, machined, rather unadorned products and interior environments of the Depression years may have seemed too reminiscent of the totalitarianism that the nation had just defeated. Whether consciously or unconsciously, Americans desired "homey" environments that would provide a sense of being in touch with historical roots. Plastics gained increasing application in furniture, fabrics, wall and floor coverings, decorative housewares, and appliances, but once again, as in the early celluloid era of plastics, imitation of more expensive natural materials became the desired end. During the late 1940s, *House Beautiful* and *Better Homes and Gardens* ran articles on "how to put plastics together in a comfortable, normal house instead of a facsimile of a Statler cocktail bar." Rather than being "by nature shiny, sleek, and a little too strange-looking for the living room," plastics could be "homey, chintzy, and comfortable." The goal became selecting furnishings that "are plastic but don't look it." And the utopian fervor that plastics had once evoked had narrowed to one minor element. Synthetic materials were "constitutionally averse to staining, scuffing, and deterioration."[58] A housewife could clean them, as everyone liked to point out, with a mere wipe of a damp cloth.[59]

Whatever their usefulness to human life and comfort, plastics in the postwar era no longer inspired any vision greater than what might be called "damp-cloth utopianism." Traveling full circle to their origin a hundred years earlier, plastics had become materials admired and used primarily for their imitative possibilities rather than for their intrinsic value as expressions of a utopian technological future. At least in their five-and-dime manifestations, plastics had also earned the unfortunate reputation of being cheap and shoddy. They were substitutes—nothing more, nothing less—and increasingly fair game for any critic who wished to find American life shallow, superficial, phony, or abstracted from reality. This image problem plagued the industry for thirty years, prompting the editor of *Modern Plastics* to complain in 1969 that "somebody must like the stuff" because consumption of plastics continued to expand even though "plastics' reputation has remained about as low as it can get."[60] In recent years, however, designers and the public have once again begun to appreciate plastics for their own unique qualities of appearance and texture. Perhaps stimulated by the successes of the space program, some people are buying furniture and accessories that frankly express and even celebrate the sleek plastic materials from which they are made.[61] Even so, although synthetics now command a prominent place in American "high-tech" environments, it is hard to imagine men and women ever again dreaming of their utopias as fabricated of plastic.

Notes

1. J. Harry DuBois and Frederick W. John, *Plastics*, sixth edition (New York: Van Nostrand Reinhold, 1981), p. v.

2. See Helen Dahlskog (ed.), *A Dictionary of Contemporary and Colloquial Usage* (Chicago: English-Language Institute of America, 1972), p. 22, for one of the first notations of this new meaning.

3. V. E. Yarsley and E. G. Couzens, "The Expanding Age of Plastics," *Science Digest* 10 (December 1941), pp. 57–59.

4. Ibid., pp. 59–60.

5. Charles M. A. Stine, as quoted in "Du Pont Launches Synthetic Silk," *Business Week*, October 29, 1938, p. 18.

6. Alexander Parkes, as quoted by Robert Friedel, *Pioneer Plastic: The Making and Selling of Celluloid* (Madison: University of Wisconsin Press, 1983), p. 8.

7. "What Man Has Joined Together . . . ," *Fortune* 13 (March 1936), p. 69.

8. Friedel, *Pioneer Plastic*, p. 89. See also pp. 29–30, 65. Other histories of the plastics industry are J. Harry DuBois, *Plastics History U.S.A.* (Boston: Cahners, 1972); M. Kaufman, *The First Century of Plastics* (London: Plastics Institute, 1963).

9. Leo Baekeland, "The Synthesis, Constitution, and Uses of Bakelite," *Journal of Industrial and Engineering Chemistry* 1 (March 1909), pp. 156–157.

10. C. F. Chandler, "Presentation Address," included in "Perkin Medal Award," *Journal of Industrial and Engineering Chemistry* 8 (February 1916), p. 181.

11. Henry Smith Williams, "The Miracle-Workers: Modern Science in the Industrial World." *Everybody's Magazine* 17 (October 1907), pp. 497–498.

12. Edwin E. Slosson, "Chemistry in Everyday Life," *Mentor* 10 (April 1922), p. 3.

13. Ibid., pp. 3–4, 7, 11–12.

14. John K. Mumford, *The Story of Bakelite* (New York: Robert L. Stillson, 1924), pp. 7, 20, 22, 46, 51.

15. "What Man Has Joined Together . . . ," p. 69.

16. "Plastics' Progress," *Business Week*, December 21, 1935, p. 17.

17. Paul T. Frankl, *Form and Re-form: A Practical Handbook of Modern Interiors* (New York: Harper, 1930), p. 163.

18. See Williams Haynes, *American Chemical Industry: The Merger Era* (New York: Van Nostrand, 1948), p. 348.

19. As quoted by Haynes, p. 349.

20. As quoted by Haynes, p. 349. For a perceptive analysis see Eleanor Gordon and Jean Nerenberg, "Everywoman's Jewelry: Early Plastics and Equality in Fashion," *Journal of Popular Culture* 13 (spring 1980), pp. 629–644.

21. "Plastics' Progress," p. 16; Haynes, *American Chemical Industry: The Merger Era*, pp. 350–351; Williams Haynes, *American Chemical Industry: Decade of New Products* (New York: Van Nostrand, 1954), pp. 330–331.

22. A. E. Buchanan, Jr., "Synthetic Houses," *Scientific American* 149 (October 1933), p. 149; Joseph Singer, *Plastics in Building* (London: Architectural Press, 1952), pp. 25–27; Haynes, *Decade of New Products*, pp. 338–339.

23. "By-Product Becomes Unbreakable Tableware," *Business Week*, October 29, 1930, p. 15; Haynes, *Merger Era*, p. 354; Haynes, *Decade of New Products*, pp. 335–336.

24. A. M. Howald, "Systematic Study Develops New Resin Molding Compound," *Chemical and Metallurgical Engineering* 38 (October 1931), pp. 583–584; James L. Rogers, "Plaskon, a New Molding Compound the Result of Planned Research," *Plastics and Molded Products* 7 (December 1931), pp. 664–665, 687; "Giant Plastic Molding Press Produces Large Weighing Scale Housings," *Iron Age* 136 (August 29, 1935), pp. 13–14; H. D. Bennett, "Pushing Back Frontiers," *Modern Plastics* 13 (September 1935), pp. 25–27, 30–32.

25. See Haynes, *Merger Era*, pp. 351–352; Haynes, *Decade of New Products*, pp. 337–338; Frank D. Morris, "Sheer Magic," *Collier's* 105 (April 13, 1940), pp. 13, 69–71.

26. Nathan George Horwitt, "Plans for Tomorrow: A Seminar in Creative Design," *Advertising Arts*, July 1934, p. 29.

27. "What Man Has Joined Together . . . ," p. 69.

28. "A Plastic a Day Keeps Depression Away," *Chemical and Metallurgical Engineering* 40 (May 1933), p. 248.

29. Peter Müller-Munk, "The Future of Product Design," *Modern Plastics* 20 (June 1943), pp. 77, 144.

30. Franklin E. Brill, "Some Hints on Molded Design," *Plastic Products* 9 (April 1933), p. 55.

31. Peter Müller-Munk, as quoted by Seymour Freedgood, "Odd Business, This Industrial Design," *Fortune* 59 (February 1959), p. 132.

32. "New Jobs for Plastics," *Business Week*, December 28, 1935, p. 17.

33. Harold Van Doren, "A Designer Speaks His Mind," *Modern Plastics*, September 1934, p. 24.

34. "Small Radios—Today and Tomorrow," *Modern Plastics* 17 (March 1940), p. 80.

35. Robert L. Davis and Ronald D. Beck, *Applied Plastic Product Design* (New York: Prentice-Hall, 1946), p. 44.

36. Raymond P. Calt, "A New Design for Industry," *Atlantic Monthly* 164 (October 1939), pp. 541–542.

37. On plastics and streamlining see also Franklin E. Brill, "What Shapes for Phenolics," *Modern Plastics* 13 (September 1935), p. 21; Franklin E. Brill and Joseph Federico, "Decorative Treatments for Molded Plastics," *Product Engineering* 8 (January 1937), pp. 23–25; Frank H. Johnson, "Designing Plastic Parts," *Product Engineering* 9 (February 1938), p. 61; John Sasso and Michael A. Brown, Jr., *Plastics in Practice: A Handbook of Product Applications* (New York: McGraw-Hill, 1945), pp. 23–24; J. Harry DuBois, "Plastics Product Design," in Henry Richardson and J. Watson Wilson (eds.), *Fundamentals of Plastics* (New York: McGraw-Hill, 1946), p. 261; John Sasso, *Plastics Handbook for Product Engineers* (New York: McGraw-Hill, 1946), pp. 342–352.

38. *Literary Digest* 112 (January 2, 1932), p. 42.

39. *Review of Reviews* 86 (November 1932), pp. 62–63.

40. Elinor Hillyer, "The Synthetics Become the Real," *Arts and Decoration* 42 (January 1935), p. 28.

41. Alden P. Armagnac, "New Feats," *Popular Science Monthly* 129 (July 1936), pp. 9–11, 109.

42. Frederick Simpich, "Chemists Make a New World," *National Geographic Magazine* 76 (November 1939), p. 601.

43. Julian P. Leggett, "The Era of Plastics," *Popular Mechanics Magazine* 73 (May 1940), pp. 130A, 658.

44. Ibid.

45. "Plastic Fords," *Time* 36 (November 11, 1940), p. 65.

46. "Ford from the Farm," *Newsweek* 18 (August 25, 1941), p. 39.

47. "Plastic Ford Unveiled," *Time* 38 (August 25, 1941), p. 63.

48. "Plastic Fords" (note 45), p. 65. On Ford's involvement with agricultural plastics see also Reynold Wik, "Henry Ford's Science and Technology for Rural America,"

Technology and Culture 3 (summer 1962), pp. 247–248; David L. Lewis, "Henry Ford's Plastic Car," *Michigan History* 56 (winter 1972), pp. 319–330.

49. Weil and Anhorn, *Plastic Horizons* (Lancaster, Pa.: Jacques Cattell Press, 1944), p. 130.

50. "In the News," *Modern Plastics* 20 (May 1943), p. 114.

51. W. Ward Jackson, "The Future of Plastics in Aviation," *Modern Plastics* 21 (January 1944), p. 176; Forrest Davis, "Airplanes, Unlimited!," *Scientific American* 161 (July 1939), p. 17. On wartime uses of plastics see Joseph L. Nicholson and George R. Leighton, "Plastics Come of Age," *Harper's Magazine* 185 (August 1942), pp. 300–301; John Delmonte, "The Postwar Role of Plastics," *Modern Plastics* 20 (April 1943), pp. 59, 142, 144, 146, 148, 150; James J. Pyle, "New Horizons in Plastics," *Science Digest* 18 (August 1945), p. 85.

52. Advertisements in *Modern Plastics* 20 (August 1943), p. 50; ibid. 20 (May 1943), p. 2.

53. Advertisement in *Modern Plastics* 20 (April 1943), p. 6.

54. "Test-Tube Marvels," *Newsweek* 21 (May 17, 1943), p. 42.

55. L. H. Woodman, "Miracles? . . . Maybe," *Scientific Monthly* 58 (June 1944), p. 421.

56. "What Does the Public Know of Plastics?," *Modern Plastics* 24 (December 1946), p. 5.

57. Examples are from Pyle, "New Horizons in Plastics," pp. 85–86; "The Buyer Is Reaching for His Crown," *Modern Plastics* 24 (February 1947), p. 5; "A 1950 Guide to the Plastics," *Fortune* 41 (May 1950), p. 111. On shoddy plastics during the immediate postwar years see also "Realities or Reveries?," *Modern Plastics* 21 (April 1944), pp. 80–81; "Propaganda . . . a Threat or a Boost," *Modern Plastics* 22 (March 1945), pp. 93–94, 198, 200; "Putting Plastics in Its Place," *Scientific American* 177 (November 1947), p. 209.

58. Quotations are from "Plastics: A Way to a Better More Carefree Life," *House Beautiful* 89 (October 1947), pp. 123, 141; Christine Holbrook and Walter Adams, "Dogs, Kids, Husbands: How to Furnish a House So They Can't Hurt It," *Better Homes and Gardens* 27 (March 1949), pp. 37–39.

59. "Damp-cloth" cleaning is mentioned in the articles cited in the previous note and in Ruth Carson, "Plastic Age," *Collier's* 120 (July 19, 1947), p. 49.

60. Sidney Gross, "Image Please," *Modern Plastics* 46 (August 1969), p. 43. See also "Let's Use the Word 'Plastics' with Pride!," *Modern Plastics* 28 (February 1959), p. 5; "What *Is* Plastics' Image, Anyway," *Modern Plastics* 47 (July 1970), pp. 66–70.

61. Contemporary plastics design is well represented by Sylvia Katz in *Plastics: Designs and Materials* (London: Studio Vista, 1978). There is apparently an awakening interest in the design history of plastics, as indicated by the recent publication of Katz's *Plastics: Common Objects, Classic Designs* (New York: Abrams, 1984) and Andrea DiNoto's *Art Plastic: Designed for Living* (New York: Abbeville, 1984), both useful illustrated surveys intended primarily for collectors.

5

Utopia Realized: The World's Fairs of the 1930s

Folke T. Kihlstedt

Most utopias are mere figments of the imagination, and most are embodied only in literature. Even in the nineteenth century, when some small utopian communities were actually built, utopian endeavors remained primarily literary. In the twentieth century, however, visionary images of the future were brought to life and offered to the public at world's fairs.

This chapter examines iconographic aspects of two American world's fairs of the 1930s—the Century of Progress Exposition (CPE), held at Chicago in 1933–34, and the New York World's Fair (NYWF) of 1939–40—and shows how these fairs projected utopian images.[1]

By *utopia* I mean what some may prefer to term *eutopia*: a good, benificent place, better in all ways than that in which its creators live. The good life of utopia was evoked by the American fairs of the 1930s as a response to the Great Depression. These fairs delivered to the large cross-section of American society that passed through their gates a buoyant, optimistic message extolling the positive consequences of science and technology for life in the future. Clearly, the organizers of the fairs had adopted from certain literary utopias of the late nineteenth century and the early twentieth century technological ideals that appealed to the masses as antidotes to the social, industrial, and commercial exigencies of the United States during the Depression.[2] These ideals were presented to the public in the themes, the buildings, and the displays of the fairs. Visitors saw a vision of a future in which democracy, capitalism, and consumerism were affirmed by science and technology. These fairs (the second in particular) equated happiness with the fulfillment of material needs and wants, as had many nineteenth-century utopias. But whereas most nineteenth-century utopias were socialist, based on cooperative production and distribution of goods, the twentieth-century fairs suggested that utopia would be attained through corporate capitalism and the individual freedom associated with it. The organizers of fairs often adapted Edward Bellamy's and H. G. Wells's widely read visions of a world of scientific progress to these ends.

The transition from the literary presentation of utopian visions to their actualization in visual form was made possible by the physicality of world's fairs as architectural creations, by the attitude toward technology in the United States in the 1930s, and by the activity of industrial designers. In the first place, these fairs existed as tangible, three-dimensional realities. Second, the technocracy crusade of the 1930s and the engineering ethos it embraced heightened the perception of utopias as feasible. To the adherents of technocracy, engineers were capable of solving all problems with dispatch, and technology was capable of creating a perfect society. More than one writer called the NYWF an engineer's utopia.[3] But as important as the engineer was, he took a back seat to a newer breed of professional: the industrial designer, who quickly became the chief promoter of a utopian future served by the products of technology.

Industrial designer Norman Bel Geddes was called in as a consultant to the CPE's architectural commission in 1929. Another designer, Walter Dorwin Teague, created the Ford exhibits at the same fair. The success of his work at Chicago earned Teague a place on the NYWF's original theme committee and on its board of design. In fact, as Teague remarked, each major exhibit at the NYWF, as well as the fair's theme, was "entrusted exclusively to Industrial Design," because "the industrial designers are supposed to understand public taste and be able to speak in the popular tongue, and because as a profession they are bound to disregard traditional forms and solutions and to think in terms of today and tomorrow."[4]

A slick practitioner of technocratic thinking, the industrial designer was an ideal synthesizer of scientific concepts for mass education and an important shaper of public taste. His was a brand-new discipline lacking the constraints that had developed naturally in the specialized training of older professions such as architecture and engineering. This lack of prior constraints allowed the industrial designer to take a fresh look at the role of science and technology in culture. He looked not with the pragmatic eye of the engineer but with the visionary gaze of the utopian. For example, Bel Geddes proposed such visionary but impractical schemes as his enormous Airliner Number 4, a tailless swept-wing monoplane with a 528-foot wingspread. For the CPE he proposed an aerial restaurant with three revolving floors cantilevered from a 278-foot service column.[5] Teague, a more cautious man by nature, still discussed design in his book with references to H. G. Wells and held to the sentiment that ideal design would usher in a golden age of peace and harmony for mankind.[6]

The Hall of Science was designated the primary building of the CPE. The organizers of NYWF went even further: Rather than relegate science and the displays of its accomplishments to a single, separate hall (and thereby "divorcing science from life," as one organizer put it), the theme committee under Teague saw science as the most powerful social force and decreed that it should permeate all the exhibits.[7] The stress on the improvement of life through science linked these fairs to the one clearly utopian mode of the twentieth century: the scientific utopia, which today's leading historians on utopian writing, F. E. and F. P. Manuel, claim is "the only form in which the utopian mode, born in a pre-industrial age, is able to survive."[8]

Earlier world's fairs, in which science had not played so great a role, had also been conceived in a utopian spirit,[9] but not until the 1930s did science and technology seem to possess the potential for the actualization of a utopian vision. Such an actualization was implicit in the theme of NYWF, "The World of Tomorrow." The displays presented this world as within technology's grasp and requiring only enlightened administration to become reality. The CPE, too, had looked forward to a scientific future of enormous changes, but that future, according to its planners, would have to await "a generation that has not yet been born."[10] The planners of the NYWF wanted to bring utopia closer. Fair president Grover Whalen remarked: "That word 'future' bothered me; I kept thinking of fortune telling . . . it suddenly occurred to me that we might call it 'The World of Tomorrow.' "[11] Grasping Whalen's sense of the immediate, Gerald Wendt, NYWF Department of Science director, proclaimed: "The tools for building the world of tomorrow are already in our hands."[12] More forcefully than Whalen, who had evoked "tomorrow" chiefly as a promotional gimmick to sell the fair, Wendt saw "tomorrow" as nothing less than the imminent dawn of a new society.

To Wendt and other organizers of the NYWF, that society would emerge in a world of total economic interdependence whose individual citizens would enjoy an unlimited range of leisure activities and personal independence. These optimistic organizers anticipated a golden age in which capitalism would generate those leisure activities through the mass production of consumer goods. A society weaned on leisure activities and surrounded by material goods was certain to appeal to the masses of fair visitors who sought to erase the memories of the Depression. The promise of leisure and abundance reflected nothing less than a struggle for survival. The fair organizers committed themselves to showing, in the midst of the Depression, how the United States could maintain democracy in the face of the growing threats of communism and fascism, redress the alienation between the individual and the community, and provide plenty for all.[13]

These concerns intensified with the spreading effects of the Depression. With unemployment up and production down, many began to think, as did Harold Loeb, a leading figure in the technocratic movement of the 1930s, that "if capitalism cannot adjust itself to its offspring, technology, a new system must be found."[14] In meeting what was perceived as a threat to democracy and capitalism, the organizers of the NYWF called for a "consumer's fair," stating: "We must demonstrate an American Way of Living."[15] The *Official Guide Book* openly claimed that "the fair exalts and glorifies democracy as a way of government and a way of life."[16]

Superficially, that way of life was embodied in the general style of the architecture at both fairs. The buildings had low, horizontal profiles with windowless, smooth exterior surfaces modulated by bright color, lights, and rounded corners. In a word, they were streamlined. Sheldon and Martha Cheney, students of art and astute observers of American culture, claimed in 1936 that "we subjectively accept the streamline as valid symbol for the contemporary life flow."[17] The streamlined architecture of the fairs carried associations with technological precision and efficiency as well as optimism about a unified and smoothly functioning future.

But streamlining made only superficial references to a technological utopia of the future. On a deeper level, the architecture of the fairs also borrowed from the descriptions to be found in certain literary utopias. One of the most influential of these literary utopias was Edward Bellamy's *Looking Backward* (1888). The world which Mr. West, Bellamy's hero, finds upon awaking into the year 2000 is a "high-tech" world of soaring skyscrapers, streets covered with transparent material, and music piped into the home. Bellamy's utopia even idealizes retirement rather than work, thus presaging the emphasis on leisure activities in the displays of the NYWF. Additionally, just as the "World of Tomorrow" was an imminent vision, so Bellamy admitted that "*Looking Backward* was written in the belief that the Golden Age lies before us and not behind us, and is not far away."[18]

Bellamy's utopian vision, which gained immediate popularity, was soon augmented by the prolific utopian writing of H. G. Wells. Although not all Wells's early utopian writing was optimistic, *A Modern Utopia* (1905) posited the ideal of "a dynamic technological society of joy and endless movement."[19] Like Bellamy's work, Wells's novel made far-reaching contributions to utopian literature and thought—among them the concept of a planetwide utopia, the importance given to science and technology, and the idea that utopia is "not static perfection but an ever evolving dynamism."[20] These same attitudes were among the guiding principles of the 1930s' fairs. Furthermore, Wells can be considered the creator of the modern idea of time travel,[21] which was the basis of two important exhibits at the NYWF: General Motors' Futurama and Westinghouse's Time Capsule.

The influence of Wells on the fairs extends beyond the relationship between time travel and these two exhibits. Even his early novel *When the Sleeper Wakes* (1899) contains many details that were brought to life in the 1930s' fairs—for example, windowless houses with central lighting and air conditioning. The hero of this novel falls into a trance in 1897 and awakes in the spring of 2100 in a technological city similar in concept and details to what was depicted in some of the dioramas at the NYWF.

The Wellsian supercity would come to symbolize the twentieth-century urban vision of a future society altered by the beneficent powers of science, technology, and mechanization,[22] and the NYWF's designers adopted this vision of the future. More than likely, their theme, "The World of Tomorrow," was inspired by Wells, even if the phrase was coined by Grover Whalen. Wells declared in 1902 and repeatedly thereafter that "tomorrow is the important and fruitful thing."[23] At the age of 72, Wells contributed the lead article to a special section the *New York Times* devoted to the fair. His article, entitled "World of Tomorrow,"[24] was overlaid with a visionary rendering showing a bird's-eye view of a monumental city of the future with streamlined buildings that seemed molded out of some synthetic substance. On the far left edge was the dark silhouette of the fair's "theme center," the Trylon and Perisphere—a subtle implication that the real world of tomorrow lay just beyond the fairgrounds.

The idea of a utopia so close in time to the fair was taken up in the text of Wells's essay. He envisaged a world in which advanced communications would erase the difference between town and country, patriotism would be obsolete, and "a collective human intelligence will be appearing and organizing itself in a collective human will."[25] Here Wells was transferring some of his own earlier utopian ideas to the utopia of the NYWF.

Some of the buildings, displays, and films of the 1930s' fairs reflected earlier literary utopias. Others were derived from earlier architectural utopias, and still others were new creations. Running through them all, however, was the image of an ordered and smoothly functioning society—an image that contrasted starkly with the faltering economy and the social dislocations of the period.

At the NYWF, the Trylon and Perisphere was at the center of an essentially circular plan defined by avenues radiating from it like spokes. A main axis running roughly from north to south culminated in the large U.S. Government building at one end and the Chrysler Motors building (bracketed by the General Motors and Ford buildings) at the other—a clear allusion to the unification by the theme center of the public and private sectors. If this axis had been continued to the south, it would have passed through the Statue of Liberty, thus connecting the fair to a symbol of the democracy underlying American capitalism. As a further development of this theme, the exterior surface of the Perisphere was bathed at night in colored lights, making it look like a planet or the earth as seen from space.[26] At times, these lights were red, white, and blue, clearly alluding to American democracy and capitalism extending over the entire planet. The Perisphere, in conjunction with the "Democracity" it housed, symbolized a future global state.

The radiating plan of the NYWF was surprisingly traditional for a fair that otherwise stressed a modernist style of architecture. The plan of the CPE, on the other hand, was unabashedly modern. It was dominated by a long, winding road, about which the buildings were grouped in conformity with the flow of traffic. It presented, in contrast with axes, courts, and vistas (the traditional planning elements that had regulated the layouts of all earlier world's fairs), an "evolving incipient roadtown."[27] Here the road was acknowledged as an important element of planning. This layout was in keeping with the windowless, horizontal, "streamlined" architecture of the CPE. The planners of the NYWF reverted to a less clearly modern, centralized layout with radiating streets for the purpose of stressing order and control. Furthermore, the particular formal arrangement they chose recalls numerous utopian city plans of the Renaissance and the Enlightenment, such as Claude-Nicolas Ledoux's City of Chaux (1773–1779).[28] The centralized layout of NYWF was even closer to that of Tommaso Campanella's Pansophic utopia, the City of the Sun (1623),[29] which had a circular temple at its center.

The architects of the NYWF's theme center, Wallace K. Harrison and Jacques Fouilhoux, were familiar with the work of the French utopian architects of the Enlightenment, if not with Campanella, and in 1933 Emil Kaufmann had published a book (*Von Ledoux bis Le Corbusier*) linking the French utopian architect with the modern

Figure 1
Claude-Nicolas Ledoux, design for caretaker's house at Maupertuis, ca. 1780. Bibliothèque Nationale, Paris.

movement.[30] To Ledoux, the spherical architectural form was an embodiment of perfection. To Harrison and Fouilhoux, it was merely a "trademark or symbol" of the fair. Yet in search of that symbol they claimed to have "ransacked every book we could find and the entire brains of our office for ideas."[31] In view of this claim, it is likely that they discovered the numerous spherical buildings of Ledoux and the other French utopians. Ledoux's spherical house for the caretaker at Maupertuis (figure 1) was illustrated in Kaufmann's book. This, or one of many other designs by the eighteenth-century French architects whose work began to be published in the wake of Kaufmann's book, is likely to have influenced the Perisphere. Another possible influence is J.-N. Sobre's spherical Temple of Immortality (1802),[32] which was intended to give the impression of floating above a pool of water. Harrison and Fouilhoux had intended the Perisphere to give the illusion of floating on a mist of water through the use of mirrors and fountains, which would have concealed its steel supports.[33]

Thus, the theme center shared a spherical form and an illusionistic intention with the utopian designs of French Enlightenment architects. At the same time, it housed a display, created by industrial designer Henry Dreyfuss and called Democracity, that alluded to some different utopian concepts. Democracity was a vast diorama of a city of the year 2039 upon which visitors looked down from two peripheral balconies within the Perisphere. Centerton, its hub, with a workday population of 250,000, had radiating streets and low buildings culminating in a central skyscraper of 100 stories. This dominant skyscraper resembled the utopian Stadtkrone (city crown) imagined in 1919 by the German expressionist architect Bruno Taut. Writing about his Stadtkrone at a time when German architects could only dream of designing such buildings, Taut claimed: "The socially directed hopes of the people find their fulfillment here in the heights."[34] Similarly, Dreyfuss's Democracity was seen as embodying "all the elements of society linked together for the common good."[35]

Surrounding Centerton were about seventy satellite towns devoted to various industries and housing both workers and management. Beyond these towns lay farms and cattle country. This layout closely resembled another urban utopia: Ebenezer Howard's Garden City (1898–1902), in which London would have become the "Centerton."[36]

The lighting of the Democracity diorama reproduced a 24-hour day in 5 1/2 minutes. The climax of this "day" was particularly dramatic: "As dusk fell, the ceiling of the globe glowed with stars. Accompanying a symphonic poem, a thousand-voice chorus came from the heavens, and from the equidistant points on the horizon came marchers representing various groups in society. The marchers increased in size, then vanished beyond drifting clouds. A blaze of polarized light was the climax."[37] The globe whose ceiling glowed with stars was an important feature of Campanella's round church in the center of his City of the Sun, not to mention some projects by the French Enlightenment architects and, more prosaic, contemporary planetaria. The heavenly chorus suggested the chorus of angels in the book of Revelation. To visitors, it must have seemed almost as if the Platonic demiurge were still at work, fashioning a universe in ideal, spherical form, or as if Democracity were a feature of the Second Coming. But the marchers who appeared from equidistant points in the heavens suggest a more specific iconographic source for Dreyfuss's diorama. As they marched, they sang the theme song of the New York World's Fair:

We're the rising tide coming from far and wide
Marching side by side on our way,
For a brave new world,
Tomorrow's world,
That we shall build today.[38]

These marchers represented the various segments of society, including laborers. They parallel the central panel of the *Ghent Altarpiece* (1432) by Jan Van Eyck, which depicts "a great multitude . . . of all nations

Figure 2
Norman Bel Geddes, City of 1960, Futurama exhibit, General Motors Building, New York World's Fair, 1939–40. General Motors

and kindreds, and people" (Revelation 7). In this well-known painting, the saints converge toward the altar of the Lamb from the four corners of the world. As they reveal the unity and the "ultimate beatitude of all believing souls," these saints define by their presence a heaven on earth.[39]

Democracity's allusions to the *Ghent Altarpiece* and to Revelation, though not indisputable, are characteristic of the eclectic syntheses that Dreyfuss and other industrial designers often attempted. Its iconographic adaptations from past sources need not have been understood for it to have stirred the fair visitor who saw it.

Even more powerful, and by far the most popular diorama at the NYWF, was a second urban scheme: the "City of 1960" in Norman Bel Geddes's Futurama (figure 2), the key attraction in General Motors' "Highways and Horizons" exhibit. The Futurama, which could accommodate 27,000 people per day on its conveyor-belt system of easy chairs wired for sound, carried visitors on a 1,600-foot, 15-minute trip. After bearing the visitor aloft to look down upon futuristic farms, bridges, and superhighways, this armchair tour entered the "City of 1960" and culminated in a close-up view of a street intersection. The model city combined widely spaced, streamlined glass skyscrapers with lower buildings. Parks took up one-third of the city's area, and on the roofs of the lower buildings were gardens. Other roofs functioned as landing pads for airplanes and autogyros. A vast system of limited-access superhighways and small roads penetrated the lower levels of the "City of 1960."

Bel Geddes, who was quite familiar with the projects of the European modernist architects, probably adapted his model city, both in form and layout, from the idealized urban plans of the highly influential Swiss-French architect Le Corbusier.[40] The "City of 1960" exhibited every major element to be found in Le Corbusier's utopian urban projects, such as his Plan Voisin (1925) and his Radiant City (1930). Le Corbusier's cities were also to have been dominated by highway systems, large park areas, and widely spaced skyscrapers. Just as Le Corbusier had called for a healthful and life-enhancing city emphasizing sun, space, and greenery, the Futurama narration prominently claimed "the City of 1960 has abundant sunshine, fresh air, fine green parkways."[41] Just as Le Corbusier organized his city into a grid of superblocks to enhance movement of traffic, the Futurama narrator talked about "breath-taking architecture—each city block a complete unit in itself" and "broad, one-way thoroughfares—space, sunshine, light and air." Just as Le Corbusier eulogized the engineer and the power of technology, claiming "What gives our dreams their daring is that they can be achieved,"[42] the Futurama was called "a vision already conceived by 1939's engineers" and "a vision of what Americans, with their magnificent resouces of men, money, materials and skills, can make of their country by 1960, if they will."[43]

The utopian aspects of the Futurama's "City of 1960" were underscored by the very first words heard by each armchair traveler: "Stange? Fantastic? Unbelievable? Remember, this is the world of 1960." Furthermore, in a clear allusion to H. G. Wells, a General Motors news release referred to the sound chairs as "time machines."[44]

Wells had an even greater influence on a third urban scheme at the NYWF: Walter Dorwin Teague's diorama for United States Steel. Teague's city was not as thoroughly developed as Bel Geddes's, and it lacked certain human amenities. It seemed to consist mainly of multilevel roadways and modernistic glass skyscrapers boldly supported by external trusses. Visitors looked down upon it from a balcony as if they were in one of its skyscrapers. In front of them, near the railing, stood a man whose dress was clearly of a future century. What was Wellsian about this was not merely the science-fiction ambience, but also the specific motif of glimpsing the city from a balcony; Graham, the hero in *When the Sleeper Wakes*, first sees the vast utopian city from a balcony. As already mentioned, Teague was familiar with Wells's writings, and his diorama is uncannily close to Wells's vision. (However, it would not do in the age of the automobile to have moving walkways for pedestrians, as in Wells's novel; consequently, Teague expanded Wells's moving walkway into a swarming thoroughfare full of vehicles.) A similar balcony view was illustrated in an early edition of *When the Sleeper Wakes*,[45] and the motif of looking down upon a city from a high balcony seems to have become an accepted way of introducing a "visitor" to an unfamiliar utopian city of the future. Hugh Ferriss also adopted the balcony parapet as the point from which his "metropolis of tomorrow" was viewed.[46]

In a sense, the individual moving chairs of GM's Futurama were equivalents to the balcony. However, they did not keep the viewer separated from the diorama; they delivered him right into it. As the chairs passed close to a typical street intersection of the "City of 1960," the narrator announced: "In a moment we will arrive on this very street intersection—to become part of this selfsame scene in the World of Tomorrow—in the wonder world of 1960—1939 is twenty years ago! ALL EYES TO THE FUTURE." The 15-minute diorama excursion prepared the viewers for their deposit into the scene (figure 3). Most were ready to suspend their disbelief. "It is a brilliant *coup de theatre*," wrote one visitor. Only the sardonic critic Buce Bliven chose to notice that the streets of 1960 were "filled with General Motors trucks and cars, definitely of 1939 vintage" rather than with the teardrop vehicles Bel Geddes had designed for the diorama.[47]

Any utopia responds to and builds from contemporary actualities, and Bel Geddes's "street intersection of tomorrow" was no exception. As a comparison of figures 3 and 4 will show, Bel Geddes developed his intersection from a proposal by architect Harvey Wiley Corbett for transforming New York City to suit the automobile. Corbett had proposed, in 1927, a step-by-step transformation that was to culminate in an "avenue of the future" with suspended roadways and automobile-pedestrian separation, much as in the GM exhibit. Only the architectural style had changed over the twelve years—a change due in great part to Bel Geddes's professional commitment to modernism and to the influence that the work of German architect Erich Mendelsohn had on Bel Geddes's architectural forms. By late 1935, Corbett had become one of the initial supporters of a New York world's fair, and it was probably with his approval that Bel Geddes "repackaged" Corbett's "avenue of the future" in a more up-to-date, modern form. Both men felt an urgent need for "an adaptation of the metropolis to the needs of traffic," to quote Corbett, who claimed that "fast movement of

Figure 3
Norman Bel Geddes, street intersection of the future (full-scale mockup), General Motors Building, New York World's Fair, 1939–40. *Architectural Record* 86 (August 1939), p. 44.

traffic of all forms . . . is essential to healthy growth."[48] For his part, Bel Geddes saw "a free-flowing movement of people and goods" as "a requirement of modern living and prosperity."[49] The response to Bel Geddes's intersection was enthusiastic, and the editors of *Life* fully expected it to be a constituent of a modern utopia.[50] It was more than a small-scale model; it was a full-scale fragment of the new reality. The intersection was an embryonic cell of a yet unborn world; a world sustained in the windowless, womblike building designed for GM by Bel Geddes and Albert Kahn; a world that General Motors would help bring into being after the war.

The housing displays of the fairs were also full-scale fragments of the future. Some of the houses clearly were utopian, such as the "House of Tomorrow" (1933) and the "Crystal House" (1934), both designed by George Fred Keck for the CPE and full of the most advanced technological devices imaginable.[51] The houses displayed at the NYWF were not as modernistic, but the House of Glass, sponsored by Pittsburgh Plate Glass, was welcomed as a catalyst for social changes. Gerald Wendt claimed that living in glass houses would turn people "outward instead of inward."[52] He also claimed that the House of

Figure 4
Harvey Wiley Corbett, avenue of the future, proposed for New York City, 1927. *Architectural Forum* 46 (March 1927), p. 207.

Glass was "one of the relatively few perfect scenes from the World of Tomorrow actually on display at the fair."[53]

The houses at the NYWF were not very exciting in architectural terms: "There they stand," lamented critic Bruce Bliven, "a motley of styles and periods including Colonial, neat as pins and looking precisely like part of the less fashionable section of Bronxville."[54] As an ensemble, however, the NYWF houses evoked the image of another utopian dream: the garden-city suburb. The fifteen houses of the Town of Tomorrow were laid out along a curving road like a greenbelt town or a fragment of a garden city. The curving road was even named Garden Way.

The Town of Tomorrow was intended to be "a greenbelt town in which air and light and space are utilized to their maximum."[55] So again, Ebenezer Howard's utopian vision of garden communities influenced the NYWF's planners. But whereas Democracity was patterned quite closely after Howard's urban layout, the houses in the Town of Tomorrow really focused on conveniences and appliances. It was their interior appointments that revealed the extent to which "consumer engineering constituted the fair's major element of social planning."[56]

Another powerful means of transporting people into a utopian future was film. In some ways, film was a more attractive and easier medium to experience than architecture, since the viewer did not have to expend any effort moving through it. Film also could captivate vast audiences more rapidly and less expensively; Bruce Bliven observed that "more people go to the cinema every week than visited The Century of Progress in its whole summer."[57] In a similar vein, critic Eugene Raskin, commenting on the NYWF two years before it was to open, noted that the problem of an exposition is "to create what is basically a spectacle, which people all over the world will pay to see, and in which commercial organizations will pay for the privilege of participating," and that therefore it was obvious that "the best way to build a World's Fair is not to build it at all, but to make a motion picture of it."[58] No one seriously considered substituting a motion picture of the NYWF, but enough exhibitors shared Bliven's and Raskin's faith in the extraordinary power of film that over 500 different films were shown at the fair.

Governments, corporations, private individuals, and nonprofit groups made or showed instructional, documentary, travelogue, newsreel, advertising, or public-relations films. Many of these were trite and uninspiring, but some took good advantage of the medium to develop the theme of "tomorrow's world." Film director Richard Griffith remarked in reference to NYWF films that motion pictures led the way in this "large-scale experiment with the propagation of ideas." He called film "a new invention still in process of development," having "more associations with the future than with the past."[59]

Film could present a more convincing vision of the future than could a diorama. Through film, one could control absolutely what one presented to the viewer. According to Richard Griffith, the documentary film "sees life in terms of change. . . . And in thus showing exactly how the future can be built, documentary provides an assurance that it *will* be built."[60] The most famous documentary shown at the NYWF was *The City*, by Henwar Rodakiewicz, Pare Lorentz, and Lewis Mumford. Presented with its selective camerawork showing the worst of Pittsburgh and the best of Greenbelt, Maryland, what viewer could reject its endorsement of decentralized garden cities?[61]

The documentary film, as Griffith noted in 1938, had become "a vivid, urgent method for developing the social attitudes of masses of people, for reconditioning their civic thinking."[62] Social and philosophical issues were addressed not only by intellectuals such as Mumford, but also by corporations, albeit in films that were intended for the purposes of relations or advertising.

RCA's film, *The Birth of an Industry*, praised the virtues of industrial development and scientific research. It then introduced television as a new art, calling it "a new service whose purpose is constructive in a world where destruction is rampant." It concluded in beneficent tones, touting RCA as "a new industry to serve man's material welfare."[63]

Less subtle was a humorous, experimental advertising film by Chrysler, *In Tune With Tomorrow*, which showed the assembly of a Plymouth car. More accurate, it showed a Plymouth assembling itself. Its humor was derived from its imagery of dancing springs and valves, a camshaft that searched out its proper place and fitted itself into the engine block, and four tires that sashayed in singing "My body is in the plant somewhere" to the tune of "My Bonnie Lies Over the Ocean." (The experimental aspects of this film were its clever animation and the fact that it was made in the newly invented 3-D process.[64]) Superficially, this magical assembly of a car was merely a clever advertisment for Plymouth. But when its narrator, Major Edward Bowes, informed the audience that "Chrysler . . . has always thought of the future," viewers got an inkling of a deeper meaning. Implicit in the Plymouth that assembled itself was the message of a future world order that would be achieved artificially through the application of automation and new technology.

An American way of life built by democracy, supported by technology, and nourished by consumer capitalism—the underlying message of the NYWF—was the inspirational basis for the Westinghouse Time Capsule. Likened to Wells's imaginary time machine, the Time Capsule was a torpedo-shaped container made of cupaloy and measuring 7 1/2 feet in length and 8 3/8 inches in diameter. Inside its inner crypt, lined with Pyrex glass and filled with inert nitrogen, were placed 35 everyday objects (including a toothbrush, a slide rule, and a Lilly Dache hat); samples of metals, alloys, textiles, and other materials; 23,000 pages of text as well as reproductions of music and art on microfilm; three reels of newsreel film; and instructions for making a movie projector and other machines. This "eight hundred pound letter to the future"[65] was deposited in a specially prepared shaft 50 feet below the Westinghouse building at the fair at noon on September 23, 1938—the moment of the autumnal equinox. The shaft was sealed on September 23, 1940, during the concluding weeks of the fair.[66]

The Time Capsule is to be opened in the year 6939. To ensure its discovery after five millennia, Westinghouse prepared 3,650 copies of a *Book of Record of the Time Capsule* and distributed these to libraries, museums, monasteries, convents, lamaseries, temples, and other repositories throughout the world. This book gives the geodetic coordinates of the shaft, offers instructions for making electromagnetic instruments to locate the capsule, provides a key to the English language of 1938, and contains messages to the future by Albert Einstein, Robert Millikan, and Thomas Mann.

When the Time Capsule was lowered into its crypt in 1938, A. W. Robertson, chairman of the board of Westinghouse, addressed those present in words that recall (at least metaphorically) Wells's utopian novel *When the Sleeper Wakes*: "May the Time Capsule sleep well. When it is awakened 5,000 years from now, may its contents be found a suitable gift to our far-off descendants."[67]

Even though it is intended for a future world peopled by our far-off descendants, the Time Capsule actually holds the keys to a utopia of 1939; it refers not to some future utopia but to its own euphorically optimistic, technology-oriented world. What the capsule houses is actually a blueprint of 1939 America. In the words of the fair's science director, Gerald Wendt, it will present "a full picture of American life in 1939 to whatever creatures may be living here in the year 6939." "If civilization has perished by then," Wendt said, "it can be rebuilt with the time capsule as a text."[68]

To Wendt, then, an ideal society already existed. This sentiment was shared by the organizers of the NYWF, the industrial designers,

Figure 5
The time capsule at the Westinghouse Building, being guided into its "Immortal Well" by A. W. Robertson and Grover A. Whalen, September 23, 1938. *S. I. Annual Report*, Smithsonian Institution.

and the executives of the corporations that erected expensive buildings in support of capitalism, consumerism, and democracy. In contast with some of its nineteenth-century literary sources, the "world of tomorrow" was not an egalitarian socialist utopia in which production and distribution were cooperatively owned, but it did resemble one in its equating of happiness with a high standard of living. It presented a modern, twentieth-century technological utopia in which abundance and leisure were the beneficent results of the reduced need for labor. Utopia may arrive by 2039, the year of Democracity with its "resurrected" society joyfully marching in the clouds. It may have arrived in 1960, the year of the Futurama and its full-scale "street intersection of tomorrow." It did arrive in 1940 for some forty chosen families that enjoyed a week's leisure in a house at the NYWF with all the latest conveniences. Or maybe it already existed, as the Time Capsule seemed to imply. Whatever the specific utopian message may have been, the organizers of the NYWF were making quasi-propagandistic use of utopian ideas and imagery to equate utopia with capitalism.

The cumulative effect of the world's-fair displays of the 1930s was to protect and bolster American consumer culture in the wake of the Depression. The organizers and the designers did this by adopting images and ideas from literary utopias of the late nineteenth century and the early twentieth century and presenting them to the public in the form of architecture, dioramas, and films.[69] These visionary displays posited a future in which scientific advancement and industrial technology acted as progressive and liberating forces—forces which promised a society of leisure and abundance, that, in fact, would be realized after World War II.

Acknowledgments

I am grateful to the Franklin and Marshall College Committee on Grants for funding my research for this essay, and to my colleagues Solomon Wank and Sue Ellen Holbrook for their critical responses and thoughtful insights.

Notes

1. I purposely neglect the San Francisco Golden Gate International Exposition of 1939 because its vision was more purely "escapist" and it was less technology-oriented than its New York counterpart.

2. On the extent and influence of the technological utopia in literature, see Howard P. Segal, *Technological Utopianism in American Culture* (University of Chicago Press, 1985). Segal, through his work on utopias, and I, through my work on world's fairs,

have arrived independently at the conclusion that the American world's fairs of the 1930s were conceived as utopias.

3. See, for instance, Helen A. Harrison (ed.), *Dawn of a New Day: The New York World's Fair, 1939/40* (New York University Press, 1980), p. 48.

4. Walter Dorwin Teague, "Building the World of Tomorrow," *Architectural Forum* 26 (April 1939), pp. 126–127. On the work of Norman Bel Geddes at the CPE, see F. T. Kihlstedt, Formal and Structural Innovations in American Exposition Architecture: 1901–1939, Ph.D. diss., Northwestern University, 1973, pp. 205–209; Norman Bel Geddes, *Horizons* (Boston: Little, Brown, 1932, pp. 159–171, 187–199. For an excellent overview of the growth and the early years of industrial design in the United States, see Jeffrey L. Meikle, *Twentieth Century Limited: Industrial Design in America, 1925–1939* (Philadelphia: Temple University Press, 1979).

5. See Bel Geddes, *Horizons*, figures 89 and 158.

6. Walter Dorwin Teague, *Design This Day* (New York: Harcourt, Brace, 1940), pp. 51, 229. On his design philosophy as inspired by the machine, see Meikle, *Twentieth Century Limited*, pp. 95, 139, 187.

7. See "Science and the New York World's Fair," *Scientific Monthly* 46 (June 1938), pp. 587–590; "The Progress of Science: Science and the New York World's Fair," ibid. 48 (May 1939), pp. 471–475. On the Hall of Science at the CPE, see Kihlstedt, Formal and Structural Innovations (note 4), p. 185.

8. Frank E. Manuel and Fritzie P. Manuel, *Utopian Thought in the Western World* (Cambridge, Mass.: Harvard University Press, 1979), p. 810.

9. The tenor with which Prince Albert greeted the Great Exhibition of 1851 as an event to unify the human race was certainly utopian. In another instance, the plan of the Paris Exposition of 1867—a single, vast iron-and-glass building of oval shape with concentric aisles—clearly was meant to symbolize the wholeness of man and the universe through utopian associations: "As in the beginning of things, on the globe of waters," stated the official exposition publication of 1867, "the divine spirit now floats on this globe of iron." The Chicago World's Fair and Columbian Exposition of 1893, with its masterful architectural layout (dubbed the Great White City), was the inspiration for William Dean Howells's utopian novel *A Traveler From Altruria* (1894), in which the Chicago fair gave the Altrurian "a foretaste of heaven" and a reminder of his own utopian land.

10. William G. Shepherd, "Fair for Tomorrow," *Collier's* 90 (September 17, 1932), p. 49.

11. Whalen, quoted in S. J. Woolf, "The Man Behind the Fair Tells How it Grew," *New York Times Magazine*, March 5, 1939, p. 3.

12. Gerald Wendt, *Science for the World of Tomorrow* (New York: Norton, 1939), p. 20.

13. This analysis of the NYWF comes from Joseph P. Cusker's excellent article, "The World of Tomorrow: Science, Culture, and Community at the New York World's Fair," in Harrison (ed.), *Dawn of a New Day*, p. 5.

14. See Harold Loeb, *Life in a Technocracy. What It Might Be Like* (New York: Viking, 1933), p. 4.

15. Michael Hare, secretary to the Fair of the Future Committee, quoted in Cusker, "The World of Tomorrow," p. 6.

16. As quoted in Kenneth W. Luckhurst, *The Story of Exhibitions* (London: Studio Publications, 1951), p. 163. Implicit in this statement is that democracy is the ap-

probation of capitalism, and that a socialist vision can be achieved through capitalism rather than through Marxism.

17. Sheldon Cheney and Martha Cheney, *Art and the Machine. An Account of Industrial Design in Twentieth-Century America* (New York: McGraw-Hill, 1936), p. 97. On streamlining in world's-fair architecture, see also Meikle, *Twentieth Century Limited*, pp. 153–210; Donald J. Bush, *The Streamlined Decade* (New York: Braziller, 1975), pp. 128–170.

18. For this analysis of Bellamy, see Manuel and Manuel, *Utopian Thought in the Western World*, pp. 762–764. Bellamy wrote the quoted statement in a postscript to the *Boston Transcript.*

19. Manuel and Manuel, *Utopian Thought in the Western World*, p. 776.

20. Mark R. Hillegas, *The Future as Nightmare: H. G. Wells and the Antiutopians* (Carbondale, Ill.: Southern Illinois University Press, 1974), pp. 66–69.

21. Ibid., p. 26.

22. The statement is Richard Gerber's, as noted in Hillegas, *The Future as Nightmare*, p. 43.

23. Antonia Vallentin, *H. G. Wells: Prophet of Our Day* (New York: Day, 1950), p. 132.

24. H. G. Wells, "World of Tomorrow," *New York Times*, World's Fair section, March 5, 1939, pp. 4–5, 61.

25. Ibid.

26. See Harrison (ed.), *Dawn of a New Day* p. 46 and color illustrations of lighting on p. 80.

27. Douglas Haskell, "Architecture 1933," *The Nation* 138 (January 24, 1934), p. 110.

28. See Emil Kaufmann, "Three Revolutionary Architects, Boullee, Ledoux, and Lequeu," *Transactions of the American Philosophical Society* 42 (October 1952), pp. 509 ff; J. C. Lemagny, *Visionary Architects. Boullee, Ledoux, Lequeu* (Houston: University of St. Thomas, 1968), pp. 109 ff.

29. See Rudolph Wittkower, *Architectural Principles in the Age of Humanism* (London: Tiranti, 1962 [1949]), p. 32; Helen Rosenau, *The Ideal City in Its Architectural Evolution* (Boston Book and Art Shop, 1959), p. 66.

30. Emil Kaufmann, *Von Ledoux bis Le Corbusier* (Vienna: Verlag Dr. Rolf Passer, 1933).

31. *Architectural Record* 81 (January 1937), p. 7; *Pencil Points* 18 (April 1937), p. 20. For further discussion of the theme center, see Kihlstedt, Formal and Structural Innovations, pp. 299–306.

32. See *Metropolitan Museum Bulletin* 26 (April 1968), p. 314.

33. The Perisphere was not sufficiently waterproof to carry off the illusion, according to *Architectural Review* 86 (August 1939), p. 62. The intended illusion was mentioned in articles published prior to the opening of the fair: *Engineering News-Record* 118 (March 18, 1937), p. 427; *New Republic*, December 1938, p. 120.

34. On Taut's Stadtkrone, see Tim Benton and Charlotte Benton, *Architecture and Design 1890–1939* (New York: Watson Guptill, 1975), p. 84.

35. "The World of Tomorrow," *Popular Mechanics* 70 (August 1938), p. 172.

36. Meikle (*Twentieth Century Limited*, pp. 190–192) makes reference to this connection. See also Robert Fishman, *Urban Utopias in the Twentieth Century: Ebenezer Howard, Frank Lloyd Wright, and Le Corbusier* (New York: Basic Books, 1977), pp. 23–88.

37. Francis E. Tyng, *Making a World's Fair* (New York: Vintage, 1958), pp. 33, 34.

38. The song, written by Al Stillman, is cited on p. 62 of Harrison (ed.), *Dawn of a New Day*. It bears some resemblance to Eugene Pottier's rousing communist song "The Internationale," of which (as Soloman Wank pointed out to me) it may have been a conscious parody.

39. See Erwin Panofsky, *Early Netherlandish Painting: Its Origins and Character*, vol. 1 (Cambridge, Mass.: Harvard University Press, 1964), p. 212.

40. On the general urban concepts of Le Corbusier, see Norma Evenson, *Le Corbusier: The Machine and the Grand Design* (New York: Braziller, 1969), pp. 7–23. Bel Geddes was quite familiar with the work of Le Corbusier; in his 1932 book *Horizons* he even juxtaposed the Rose Window of Rheims Cathedral and a Lycoming airplane engine (pp. 276–277) in the sort of visual analogy Le Corbusier had first made in his book *Towards a New Architecture* (1923; English edition 1927).

41. General Motors Corp., *Futurama* (1940).

42. See Fishman, *Urban Utopias in the Twentieth Century*, p. 20.

43. "Life Goes to the Futurama," *Life* 6, June 5, 1939, p. 81.

44. General Motors Corporation news release, May 30, 1939, pamphlet file, General Motors Library, Warren, Michigan.

45. For the illustration, see Ulrich Conrads and Hans G. Sperlich, *Fantastic Architecture* (London: Architectural Press, 1963), pp. 91, 174.

46. See Hugh Ferriss, *The Metropolis of Tomorrow* (New York: Ives Washburn, 1929), p. 109.

47. See "New York World's Fair,"*Architectural Review* 86 (August 1939), p. 81; Bruce Bliven, Jr., "Gone Tomorrow," *New Republic* 99 (May 17, 1939), p. 40.

48. See Harvey Wiley Corbett, "The Problem of Traffic Congestion, and a Solution," *Architectural Forum* 46 (March 1927), pp. 204, 201.

49. Norman Bel Geddes, *Magic Motorways* (New York: Random House, 1940), p. 10.

50. "Life Goes to the Futurama," p. 83.

51. See chapter 7 of this volume.

52. Wendt, *Science for the World of Tomorrow*, p. 160.

53. Ibid., p. 163.

54. Bliven, "Gone Tomorrow," p. 42.

55. Bruce Bliven, Jr., "Fair Tomorrow," *New Republic* 97 (December 7, 1938), p. 121.

56. Meikle, *Twentieth Century Limited*, p. 198.

57. Bruce Bliven, "A Century Treadmill. A Few Funeral Flowers for the Chicago Fair," *New Republic* 77 (November 15, 1933), p. 11.

58. Eugene Raskin, "Fairer than Fair," *Pencil Points* 18 (February 1937), p. 92.

59. Richard Griffith, "Films at the Fair," *Films* 1 (November 1939), p. 61. For a listing and summary of most of the films, see manuscript by Richard Griffith, Films of the World's Fair 1939, prepared for American Film Center, March 1940; a copy is in the Princeton University Library's Department of Rare Books and Special Collections.

60. Griffith, "Films at the Fair," p. 61.

61. For some contemporary responses to *The City*, see *Life* 6 (June 5, 1939), pp. 64–65; *Architectural Review* 86 (August 1939), p. 94; *Free America* 3 (August 1939), pp. 18–19; ibid. 3 (October 1939), p. 16.

62. Howard Gillette, Jr., "Films as Artifact. *The City* (1939)," *American Studies* 18, no. 2 (1977), pp. 72–73.

63. These and the following quotations and descriptions of NYWF films come from notes that I took at a special showing of the films at the Queens Museum. For the opportunity to view these films I am especially grateful to Deborah Silverman.

64. The visitor to the Chrysler building was given polarized glasses made to look like the front end of a 1939 Plymouth.

65. David S. Youngholm, "The Time Capsule," *Science* 92 (October 4, 1940), pp. 301–302. For a complete list of the articles in the time capsule see G. Edward Pendray, "The Story of the Time Capsule," Annual Report of the Smithsonian Institution, 1939, Smithsonian Institution publication 3555, 1940, pp. 533–544.

66. This was the first time capsule. Westinghouse made a second and buried it at the 1964–65 New York World's Fair. Another capsule was sealed at Expo 67 in Montreal, to be opened in 2067, and another was sealed at Expo 70 in Osaka, to be opened in 6970.

67. See Pendray, "Story of the Time Capsule," p. 539.

68. Wendt, Science for the World of Tommorrow," p. 117.

69. It would seem that these world's fairs, as optimistic utopian visions, were an antidote to the prevailing dystopian mode in literature. The contemporary works of Yevgeny Zamiatin and Aldous Huxley were antagonistic toward consumerism and skeptical of technology. Thus, one could say that the world's fairs of the 1930s replaced the literary utopian genre of the late nineteenth century while continuing its optimistic view of the future and reaching larger segments of the public.

6

The Technological Utopians

Howard P. Segal

The period from about 1880 to 1930 was marked by a proliferation of movements for social reform in the United States. The Populist and Progressive crusades—the efforts of William Jennings Bryan, Theodore Roosevelt, the Muckraker journalists, and many others to improve the lot of poorer, less powerful Americans—are well known. But there was another, concurrent movement for change that has received far less attention: technological utopianism. The utopian crusaders never organized formally; their tools were literary and reached far smaller audiences than the speeches, tracts, and political campaigns of the other movements; and the changes they sought were more comprehensive than the political and economic reforms advocated by the Populists and the Progressives. For all that, they are as much an expression of the period and of its vision of the future as their better-known and more successful compatriots.

Technological utopianism derived from the belief in technology—conceived as more than tools and machines alone—as the means of achieving a "perfect" society in the near future. Such a society, moreover, would not only be the culmination of the introduction of new tools and machines; it would also be modeled on those tools and machines in its institutions, values, and culture.

Between 1883 and 1933, twenty-five individuals published works envisioning the United States as a technological utopia. The visions differed only in minor details and may safely be treated as one collective vision, rather like a Weberian ideal type. The first of these works was John Macnie's *The Diothas; Or, A Far Look Ahead*; the last was Harold Loeb's *Life in a Technocracy: What It Might Be Like*. To be sure, visions of the United States as a technological utopia antedate Macnie's and postdate Loeb's—the earliest known one is John Adolphus Etzler's *The Paradise Within the Reach of All Men* (1833), and among the latest is

Buckminster Fuller's *Utopia or Oblivion* (1969)—but this half-century was the heyday of technological utopianism in America, at least in literary form.[1]

With the grand exception of Edward Bellamy, whose *Looking Backward* (1888) quickly became an American classic, most of the technological utopians lived and wrote in obscurity. (Several did achieve prominence in their everyday callings as businessmen or professionals, among them the carriagemaker Chauncey Thomas, the inventor and manufacturer King Camp Gillette, the civil engineer George Morison, and the mechanical engineer Robert Thurston.) A few of the technological utopians knew one another personally, and a few more knew of one another's writings, but most of them apparently worked alone. Only in the 1930s, with the brief flourishing of the Technocracy crusade, did technological utopianism become organized at all.

Because the technological utopians were not members of an organized movement and were generally obscure, it has not been easy to learn very much about their backgrounds. Nevertheless, basic biographical information on most of them is now available, and this information suggests that the technological utopians were not the marginal, alienated, disaffected figures one might expect but, on the contrary, successful and well-integrated Americans.[2] Occupationally, most appear to have been what we would call upwardly mobile, in some measure because they were part of the ever more prestigious field of technology. That the technological utopians were mainstream rather than peripheral members of society suggests that, as I shall argue, technological utopianism was a movement not of revolt but of its antithesis—a movement seeking to alter the speed with which American society was moving but not the direction.

What motivated these particular people to write the works they did? That they shared the then pervasive American faith in progress through technology is obvious, but what inspired them to become outright utopians is not. The biographical information obtained is hardly illuminating, but it is clear that the technological utopians ought not to be dismissed as crackpots or dreamers, as utopians throughout history so often have been. Their works were intended to improve but not to undermine contemporary American society. Yet, because relatively few Americans became utopians of any variety, the technological utopians ought to be accorded modest recognition exactly because they were somewhat unconventional. To treat them simply as pale reflections of more conventional reformers is to miss their significance as utopians.[3]

Even as utopians, however, the technological utopians (save Bellamy) must surely have suffered at least some disappointment at their failure to reach a wide audience, much less effect genuine change. Why Bellamy attained such fame and influence when his fellow technological utopians endured such obscurity is not easy to explain. Perhaps it was simply that he wrote better than they. Or perhaps their works were dismissed as mere imitations of or sequels to *Looking Backward.* A third and deeper explanation, which precludes neither of the other two, is that *Looking Backward*'s emphasis on cooperation and community as well as on technological advance provided a more balanced and appealing vision than the strictly materialist emphasis of nearly all the other works. Whatever the appropriate explanations, the works of those who followed Bellamy—as well as his own sequel, *Equality* (1897), a purer example of social engineering—never aroused the same enthusiasm as *Looking Backward.*

Although most of these twenty-five visionaries and their works were obscure, the values they expressed were far from obscure and had wide-ranging influence. More clearly, more methodically, and more intensely than any other group, the technological utopians espoused positions that a growing number (even a majority) of Americans during these 50 years were coming to take for granted, or wanted to: the belief in the inevitability of progress and the belief that progress was precisely technological progress. Earlier generations of Americans (and Europeans) had, of course, expressed and often acted upon these same beliefs. Technology, in rhetoric if not necessarily in reality, has invariably been deemed the foremost means of transforming the prospect of America as utopia from the "impossible" to the "possible" and, in turn, to the "probable." One can, in fact, discern an American tradition of visions of various kinds—from experimental communities to written works to world's fairs—based upon a view of technology as panacea. But no prior set of visionaries had taken those beliefs as zealously or as literally as this particular group.

Furthermore, in the face of both their and their fellow Americans' considerable concerns about the present, the technological utopians took these convictions to their logical finale: the equation of advancing technology with utopia itself. Other Americans, including such technological luminaries as Thomas Edison and Henry Ford, might equate technological progress with social progress, but they did not seek, much less expect, utopia as the final fruit of their achievements, as the technological utopians did. The historical significance of the technological utopians, then, lies in the form and the content of their visions, in their confidence in the accuracy of these visions, and in the relationship of those visions to the particular cultural context—and crises—of the years between roughly 1880 and 1930.

The visions of technological utopia were put forth in articles, addresses, tracts, short stories, and novels. Several utopians wrote more than one work, and some wrote both fiction and nonfiction. There is no qualitative distinction, however, between the fictional and the nonfictional works. All envision a similar kind of utopia, and it is especially revealing that those persons who used both genres altered none of the substance of their vision in moving from one to the other. As Kenneth Roemer observes about many of the 154 utopian and anti-utopian works of the period 1888–1900 that he studied, "the fictional form was only a sugarcoating for the authors' realistic blueprints for the future."[4] Although the amount of space and detail allocated to each dimension of a utopian society differs with each work, all treat to some degree a utopia's founding, its natural and man-made settings, its technological marvels, its institutions and values, and the personal habits and relationships of its citizens. The composite description of technological utopia presented next will make this evident.[5]

The technological utopians felt certain that they perceived quite specifically the directions in which the United States was heading, and they aimed at accurate prediction of the future rather than idle visions of a world someday, somewhere. Moreover, the world they foresaw represented no qualitative break with the existing world; many of the technological changes they predicted were by 1883 already being discussed or even developed. Insofar as the technological utopians merely extrapolated from the present to the future rather than sharply distinguish the future from the present, they may be called conservative utopians.

The utopians were not oblivious to the problems technological advance might cause, such as unemployment or boredom. They simply were confident that advancing technology held the solution to those problems and to other, chronic problems, including scarcity, hunger, disease, and war. In addition, they assumed that technology would solve the psychological problems that were increasingly worrisome, such as aggression, crowding, rudeness, and social disorder. The growth and expansion of technology they therefore equated with the coming of utopia; utopia itself they equated with a society run by (and, in a sense, for) technology. Technology seemed a far sturdier and more efficacious instrument of progress than the various panaceas proposed by their fellow utopians: taxation, socialism, religion, communitarianism, or revolution.

The technological utopians were even willing to specify utopia's time and place of arrival: usually within the next 100 years and within the existing boundaries of the United States. Yet the means by which technological utopia was to come into existence were rarely specified. Virtually every technological utopia emerges as the final stage of some vague, evolutionary scheme marked by few clear events. "It took many, many years of strife and turmoil," recounts a character in a typical work as he explains the process, "to bring about the present condition of social evolution."[6] Clearly, though, the utopian age is an age beyond history. Change will cease, because perfection will have been achieved. One citizen of technological utopia proclaims: "Eternity is here. We are living in the midst of it."[7] Another announces: "Heaven will be on earth."[8]

Despite its basis in modern technology, technological utopia was not to be a mass of sooting smokestacks, clanging machines, and teeming streets. The dirt, noise, and chaos that invariably accompanied industrialization in the real world were to give way in the future to perfect cleanliness, efficiency, quiet, and harmony. Technology, like fire, would be domesticated. Boasts one inhabitant of utopia:

> Our sanitary arrangements and lavatories are of the best, and easily accessible; our roads are well paved; smoke, cinders, and ashes are unknown because electricity is used now for all purposes for which formerly fires had to be built; our buildings and furniture, made of lacquered aluminum and glass, are cleansed by delicately constructed machinery that operates automatically. The very germs of unclean matter are removed by the most powerful of disinfectants, electrified water, that is sprayed over our walls, and penetrates into every crack and crevice.[9]

With the taming of technology was to come the taming of nature. Wind, water, and other natural resources were to be subdued and harnessed—above all, in the form of clean, quiet, powerful electricity. The mastery of nature was regarded as the fulfillment of man's destiny and his elevation to a status only slightly less than that of omnipotence. Chauncey Thomas announced: "We give Nature and her vast forces an opportunity to work for us!"[10] D. L. Stump's novel *From World to World* is dedicated to "the proposition that the earth was made to live upon, and that all its elements were created to sustain life, and for no other purpose!"[11]

The envisioned domestication of both technology and nature would lead to a resolution of the allegedly permanent tension between the industrial and the agrarian order, or the machine and the garden, which Leo Marx has said lies at the heart of the American experience.[12]

The resolution would be achieved by the modernization rather than the abandonment of the garden, by its being transported out of the wilderness and relocated in a city now itself transformed from a lethal chaos to a healthy order. The new industrialized garden would not take the form of Marx's "middle landscape" (the Jeffersonian ideal of an agrarian yet technologically proficient yeoman republic). Rather, it would take the form of a series of what we have come to call megalopolises—massive combinations of urban and suburban tracts—which would embrace practically all of utopia. The following description is representative: "All cities are now circular in form the radii of which is one hundred miles, with an approximate circumference of seven hundred miles. . . . In all there are about twenty cities. . . . Indeed, by the all potent power of electricity, man is now able to convert an entire continent into a tropical garden at his pleasure."[13] Notwithstanding their size, utopian megalopolises would enable millions to live, learn, work, and play in perfect contentment, ever free of dirt, noise, chaos, want, and insecurity.

The most immediately striking features of each megalopolis would be the buildings, numbering in the hundreds. Of various materials, shapes, and sizes, they nevertheless would fall into clear and neat patterns, with the tallest, sleekest structures concentrated in the center and the shortest and squattest along the periphery. Broad sidewalks and streets and wide plazas and parks would border each building, even in the dense downtown areas. The cold, canyonlike atmosphere of many actual skyscraper cities would thereby be avoided. "As far as I could see," recounts a visitor to utopia, "colossal, square, skillfully decorated marble palaces, built in various styles of architecture."[14] "Can you imagine the endless beauty of a conception like this," asks another, "a city with its thirty-six thousand buildings each a perfectly distinct and complete design, . . . each building and avenue surrounded and bordered by an everchanging beauty in flowers and foliage?"[15] Competition between vehicular and pedestrian traffic would be overcome through judicious placement of highways, walkways, bridges, and tunnels.

In their design, though not in their size or scale, suburban areas would resemble their urban counterparts, with which they would be tightly integrated. As a traveler to utopia remembers it, "We had left the train at what appeared to be a small village. Yet nowhere was to be seen any trace of that pervading lack of neatness and finish which, in our day, usually characterizes the country. . . . The buildings visible, though inferior in size to those of the city, were so solidly constructed, and of similar materials."[16] What farmlands would remain would also

Figure 1
An illustration from Herman Brinsmade's *Utopia Achieved* (1912).

Figure 2
The cover of Edgar Chambless's *Roadtown* (1910).

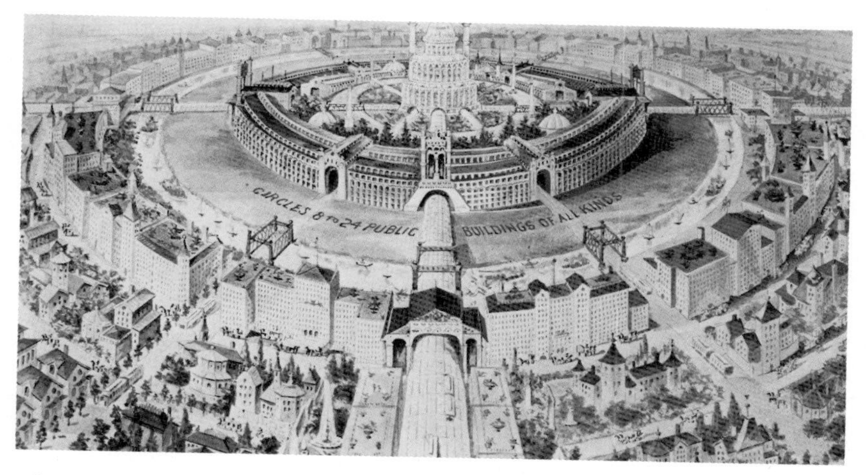

Figure 3
Sketch of the center of New Era Model city, from Charles Caryl's *New Era* (1897). Caryl depicts one of technological utopia's metropolitan centers, the equivalent of the modern megalopolis.

be assiduously organized and linked to urban regions. Agriculture, states one utopian, "is an integral part of the producing machine, and the producing machine must be a unit." Moreover, "agriculture, like industry, when reduced to its lowest terms, is a problem of applied mathematics and engineering."[17]

Connecting all sectors of the technological utopia would be superbly efficient transportation and communication systems, powered almost exclusively by electricity. These systems would enable the widely dispersed citizens to live and work wherever they might choose. As one of them puts it, "We have practically eliminated distances."[18] The specific means of transportation would include automobiles, trains, subways, ships, airplanes, even moving sidewalks. The means of communication would include pneumatic mail tubes, telephones, telegraphs, radios, and mechanically composed newspapers. In most cases the devices mentioned in the utopian works bear some resemblance to those in use at the time of writing, if such devices existed at all; however, they bear fuller resemblance to those in use today.

Technological advances inundate the homes and offices of the utopias. Electric clotheswashers and dryers, dishwashers, refrigerators, ranges, vacuum cleaners, garbage disposals, air conditioners, phonographs, razors, and haircutters fill every home. (Again, some of the devices were unknown at the time of writing but are taken for granted today.) Underground pneumatic tubes from centralized distributors would supply each home with all needed goods, from food to furniture. "It is," says one citizen, "like a gigantic mill, into the hopper of which goods are being constantly poured by the trainload and shipload, to issue at the other end in packages of pounds and ounces, yards and inches, pints and gallons, corresponding to the infinitely complex personal needs of half a million people."[19] From these various inventions comes virtual freedom from "all the annoyances of housekeeping"[20] and "everything for comfort, economy, convenience, and freedom from care that a Corporate Intelligence could think of."[21]

Curiously, the technological devices that perfect working conditions are described in much less detail—perhaps because they were more familiar to readers, even if most of them were not industrial workers. The following lines are typical: "I need not tell my readers what the great mills are in these days—lofty, airy halls, walled with beautiful designs in tiles and metal, furnished like palaces, with every convenience, the machinery running almost noiselessly, and every incident of the work that might be offensive to any sense reduced by ingenious devices to the minimum."[22] Nevertheless, automated machinery would cut the time spent working and make that time more enjoyable.

Utopia's climate would be pleasant and nearly uniform—another achievement of technology. Excessively hot regions would be cooled, excessively cool ones warmed; wet climates would be made drier and dry ones wetter. How? By the use of vaguely described, enormously powerful tools and machines with which the utopians would dredge, rechannel, and even create rivers and lakes, irrigate deserts, heat the soil, flatten mountains, clear land, and erect huge domes to capture and preserve sunlight. "We have absolute control of the weather," declares one utopian.[23]

The ethos of technology would shape the values as well as the physical dimensions of utopia. The inhabitants not only prize the virtues of technology but also strive to be as efficient as tools and machines. "The human machine," boasts one, "the greatest of all, has been . . . thoroughly understood and developed to its highest efficiency, something never previously done."[24] Another sees his fellow workers and managers as happy "cogs in the machine, acting in response to the will of a corporate mind as fingers move and write at the direction of the brain" and yearns to have "millions of individuals organized and moving like parts of a wonderful mechanism, from one field of production to another."[25] Even in their leisure time, utopians devote themselves not to rest and relaxation but to various forms of self-improvement, cultural, physical, and professional.

In imitating the tools and machines he has invented, man finds fulfillment in utopia. "We teach that labor is necessary and honorable, that idleness is robbery and a disgrace," proclaims a prominent citizen in Henry Olerich's *Cityless and Countryless World.*[26] A character in Thomas's *Crystal Button* declares: "We value time as our first of all boons—it is our life—and we count every day another opportunity freighted with duties that we take pleasure in performing."[27] Also through their characters, these same authors in fact equate work with play: "Willing work, in fields fitted to the capacity of the worker, is of itself one of the highest forms of pleasure."[28] "Work . . . gradually changes into play."[29] In short, as a third author puts it directly, "This is the age of work."[30]

The zeal for efficiency—broadly conceived as the avoidance of inefficiency or waste as much as the saving of time and energy—would shape industry, government, and education. Technical subjects, especially the sciences and the vocations, would constitute the bulk of the educational curriculum. The following pronouncement, from Gillette, is representative: "The child under the People's Corporation will have such an opportunity for education as almost no child has today. . . . The Corporation will want its children to learn its business; the miracle of scientific production; the fairy tale of flour; the romance of rubber; the wonder of wool and silk. The child will get his education in the midst of production."[31] The situation in *Roadtown*, a 1910 utopia, is similar: "Instead of college . . . there will be an industrial university. . . . Pounding literature into the head of a natural born mechanic is both economic and mental waste. The universal query in Roadtown will not be what does he know, but what can he do."[32] Nontechnical subjects would not be ignored, but they would be included only insofar as they contributed to technical knowledge: "The acquisition of the knowledge to be obtained from books, though by no means neglected, we regard as the least important branch of education."[33] "The intent of our present educational system is to prepare the individual to take some position in the industrial machine."[34] "Knowledge has become power."[35]

Upon the completion of their formal education, inhabitants of utopia invariably enter the "industrial army," the very title of which bespeaks order, discipline, organization, and therefore efficiency. As a utopian typically envisions it, "All the world is falling into line, and the whole world-wide army is moving in concert if not under a single generalship. . . . [T]he captains of industry [command] . . . mighty armies overspreading all the fields of production of a whole country, even of many countries."[36] In this civilian army all citizens serve for 20–40 years, depending upon their needs and society's. Their education amply prepares them for this service. All have technical skills, manual or mental. "Nearly every calling . . . is industrial."[37] Those with the greatest technical expertise head the industrial army. Merit alone determines status: "The principle on which our industrial army is organized is that a man's natural endowments, mental and physical, determine what he can work at most profitably to the nation and most satisfactorily to himself."[38] Since the highest merit accrues to those with the greatest technical knowledge, the leaders of the industrial army might well be viewed as the technological equivalents of Plato's philosopher-kings.

By being efficient, the citizens of utopia garner both public and private esteem but reap few tangible rewards. Hefty taxes preclude the accumulation of great wealth, and the availability of all goods and services at modest cost stifles the impetus for its accumulation. To a capitalist, the absence of any material rewards for efficiency would seem to rob work and service of any incentive. But the utopians (who, in the characteristic fashion of American reformers, were not socialists) did not envision any such threat—a fact that underscores the spell that efficiency for its own sake had cast over them.

In utopia, efficiency would govern government as thoroughly as it would education and industry. The same technical experts who would run the industrial army would run the government, because expertise, not popularity, is what utopian government would require. "Administration, in a technocracy," one utopian explains, "has to do with material factors which are subject to measurement. Therefore, popular voting can be largely dispensed with. It is stupid deciding an issue by vote or opinion when a yardstick can be used."[39] Many utopias have no politicians at all. In those that retain such a class, politicians are only figureheads.

Because technicians rather than politicians would run the utopian government, it would be technical rather than political in nature. Since the basic laws and institutions of society would have been fixed, no legal, political, or ideological tasks would remain, and therefore no lawyers, politicians, or "ideologues" would be needed. Only technical issues—which is to say, issues of efficiency—would remain, and only technicians would be needed to deal with them. In the words of one utopian, "The world owes all . . . to the inventor, to the mechanic, to the man of science. . . ."[40]

Not even culture and religion would escape the tentacles of technology and efficiency. The state would support culture, but culture would support the state and its values. "In the past," rhapsodized one technological highbrow, Robert H. Thurston, "the arts have led; in the future we shall see science leading and directing every development of the arts."[41] Some utopian writers went even further, foreseeing a world in which culture not only sanctions efficiency but is efficiency itself; in which "the greatest beauty" is that which is most "compatible with usefulness."[42] Culture, in the form of lectures, concerts, and radio, would sustain not an elite but the entire population, its mass appeal attesting still more to the utopian drive for efficiency. Religion too would promote efficiency, by upholding the value on which efficiency rests: cooperation. Indeed, religion in utopia was to consist of little more than a demythologized belief in brotherhood and the love of man. Gone would be churches, rituals, creeds, ministers, and concerns for transcendent matters such as the afterlife. Technology and science would be the gods, their deification based mainly on their efficiency. One author states, succinctly, "Real science is true religion."[43] Another labels the discoveries of science the modern version of "revelation and prophecy."[44] Another unabashedly declares engineers the "new priests" of the new society.[45]

Efficiency is not the only machinelike virtue in utopia. Self-control is equally important. Utopian man's control over himself mirrors technology's control over his environment. John Macnie writes of his utopians: "If required to state the pervading characteristic of the manners of these people, I should say self-control. In proportion as man had become master of nature, it had become needful to become master of himself."[46]

Self-control is displayed in a variety of ways. Utopian citizens conform in dress, in length of hair, in choice of food, and in the avoidance of smoking and drinking. More important, family relations are uniform: the father rules the household, and the mother and children submit. Only those citizens judged healthy in body, mind, and morals may marry and bear children, lest the others—a small minority, to be sure—perpetuate undesirables. Strict limits are imposed on the number of children permitted each couple, and divorce is forbidden except in cases of extreme incompatibility. Displays of emotion are rare, and personal relationships are ordinarily formal and distant—in other words, controlled.

Not even death warrants a release of emotion. Mourning is brief and restrained, and bodies are cleanly and efficiently cremated. Death itself is accepted as a natural phenomenon, the fundamental concern being the improvement of this life. A teacher in Solomon Schindler's *Young West* tells his young students: "Why, after all, should we trouble our minds in regard to a past that lies so far behind us. It is neither the past nor the future that should give us concern, it is the present.

Here is this beautiful world, there is the span of life, granted to us to enjoy, and here is the work by which to make this life pleasant for ourselves and others."[47]

This, then, is the composite late-nineteenth-century and early-twentieth-century vision of America as a technological utopia. The writings that express this vision constitute a unique source of information about and insight into the times in which they were written.[48] And because they constituted predictions of our own times, they deserve our further attention, if only to measure the gap between prophecy and fulfillment.

Some modern critics have deemed technological and social progress antithetical. To them, the technological progress once hailed by millions as the panacea for mankind's problems has not only failed to solve those problems but has itself become a major problem.

Clearly, neither this position nor the equation of advancing technology with utopia is wholly correct. Often one's attitude depends on the particular technology being considered—for example, household appliances or military weapons. What should be noted, however, is the irony of the fact that centuries of utopian dreams have, for many, been translated into "dystopian" nightmares.

It was in the early 1930s, in the depths of the Great Depression, that this irony first became painfully apparent to significant numbers of Americans. The latest technological marvels of industry and agriculture lay unused or underused, and technological unemployment was becoming a widespread source of concern. The feeling among many that the "Great Engineer," Herbert Hoover, had utterly failed to prevent or to stop the unprecedented economic upheaval further soured many Americans on purely technological solutions to economic, social, cultural, and political problems. In this context, it is not surprising that the last of the technological-utopian works considered here, Loeb's *Life in a Technocracy* (1933), clearly recognized that technological advances alone would not lead to a genuine utopia and that nontechnological factors, such as art and recreation, would have to be included in any satisfactory scheme.[49]

The 1930s also witnessed an extraordinary final outpouring of faith in technological progress.[50] Howard Scott's Technocracy crusade was the first organized expression of technological utopianism.[51] Like the technological utopians, the Technocrats confidently assumed that technology could bring about a far better if not an ideal society. For both, utopianism had shifted from the "possible" to the "probable." Unlike the technological utopians, however, the Technocrats assumed that a better society could be fashioned in the near future—in the next 30 or 40 rather than the next 100 years—and by groups of citizens working together rather than by isolated writers.

The world's fairs of the 1930s manifested a further shifting of the idea of utopia toward the realm of the probable. What made the American fairs of that decade distinctive was their unprecedented focus on mankind's ability to shape the future. The World of Tomorrow (the New York fair of 1939–40) even projected a technological utopia that would come about in the United States in the very near future—by 1960, to be exact.[52] It thus moved the timetable for the achievement of utopia up several decades from the century or more that was common in the utopian writings. The major architects of the World of Tomorrow, the pioneering industrial designers Walter Dorwin Teague, Henry Dreyfuss, Raymond Loewy, and Norman Bel Geddes, assumed that they could literally design utopia.[53] Just as the writings of the earlier technological utopians had brought together on paper sundry contemporary devices and ideas that had often not yet been fully developed, the fairs of 1930s gathered the devices of their day.

Although the planners of the fairs, the Technocrats, and the utopian novelists had differing agendas and held somewhat different views of the future, they all shared a belief in the capacity of technology to perfect society. That faith—a neglected strain of American culture—provides an ideal model of society against which one can measure the gaps between past and present and thus between present and future. In exploring the sources of those gaps one gains an appreciation of the difficulty of making any serious predictions.[54]

In recent years, shallow and heavily technocratic forecasts, such as those of Herman Kahn, have had to share the spotlight (if not the contracts) with the more profound, more humane visions of "alternative futures" put forth by Hazel Henderson and others.[55] Simplistic extrapolations such as Kahn's have almost invariably proved inaccurate. The field of technology assessment has, for similar reasons, had to make room in recent years for the more complex perspectives of environmental-impact assessment and retrospective technology assessment.[56] The persistence of simple extrapolations reflects the desperate but understandable search for easy answers to complex questions. This accounts for the extraordinary popularity of the muddled, unoriginal, and ahistorical prophecies of an Alvin Toffler.

In view of the manifold uncontrollable dimensions of life, the future of any but the most static society will never be reducible to a model, either in literary works or in exhibitions or in computer programs. Still, the efforts of the technological utopians and the others who have peered into the future do have value. They are vehicles of social criticism that reflect the perceived problems of their day and the possible solutions to those problems. Although such works only rarely change society directly, they can and sometimes do capture the imagination and influence the perceptions of those who have the power to effect change.

Notes

1. On these visionaries and their works, see, besides the various citations below, the appendix to my book *Technological Utopianism in American Culture* (University of Chicago Press, 1985). By contrast, I have found only three American technological utopians for the years before 1883: Etzler, Thomas Ewbank, and Mary Griffith. On their backgrounds and works, see pp. 77, 88–91, and 173–174 of the above-mentioned book.

2. See Segal, *Technological Utopianism in American Culture*, chapter 3 and appendix.

3. It is a truism that utopianism of whatever kind represents the projection of the visionary's hopes or fears or both. Yet it is imperative to avoid reducing the technological utopians' individual or collective vision to their individual or collective personalities and so warping or ignoring its content and its relationship to the "real world" in which they lived and wrote—and which they hoped to perfect. Whether the utopians' lives were happy or unhappy or a mixture of both, their writings must finally stand apart, as things in and of themselves. I have tried to follow this approach in my article "*Young West*: The Psyche of Technological Utopianism," *Extrapolation* 19 (December 1977), pp. 50–58.

4. Kenneth M. Roemer, *The Obsolete Necessity: America in Utopian Writings, 1888–1900* (Kent, Ohio: Kent State University Press, 1976), p. 3.

5. For elaboration on the description of technological utopia presented next, see my *Technological Utopianism in American Culture*, chapter 2. That chapter contains more quotations and more page citations from the other technological utopian writings for each of the points made below; the quotations used here are merely representative of the technological-utopian genre I have examined.

6. Thomas Kirwan (pseud. William Wonder), *Reciprocity (Social and Economic) in the Thirtieth Century, the Coming Cooperative Age; A Forecast of the World's Future* (New York: Cochrane, 1909), p. 115.

7. Charles W. Wooldridge, *Perfecting the Earth: A Piece of Possible History* (Cleveland: Utopia, 1902), p. 325.

8. King Camp Gillette, *World Corporation* (Boston: New England News, 1910), p. 240.

9. Solomon Schindler, *Young West: A Sequel to Edward Bellamy's Celebrated Novel "Looking Backward"* (Boston: Arena, 1894), p. 45.

10. Chauncey Thomas, *The Crystal Button: or, Adventures of Paul Prognosis in the Forty-Ninth Century*, ed. George Houghton (Boston and New York: Houghton Mifflin, 1891), p. 96.

11. D. L. Stump, *From World to World* (Asbury, Mo.: World to World, 1896), p. 2.

12. See Leo Marx, *The Machine in the Garden: Technology and the Pastoral Ideal in America* (New York: Oxford University Press, 1964). See also my "Leo Marx's 'Middle Landscape': A Critique, A Revision, and An Appreciation," *Reviews in American History* 5 (March 1977), pp. 137–150.

13. Albert W. Howard (pseud. M. Auburré Hovarrè), *The Milltillionaire* (Boston: no publisher, 1895), p. 9.

14. Paul Devinne, *The Day of Prosperity: A Vision of the Century to Come* (New York: Dillingham, 1902), pp. 56–57.

15. King Camp Gillette, *The Human Drift* (Boston: New Era, 1894), p. 97.

16. John Macnie (pseud. Ismar Thiusen), *The Diothas; Or, A Far Look Ahead* (New York: Putnam, 1883), p. 31.

17. King Camp Gillette, *The People's Corporation* (Boston: Ball, 1924), pp. 125, 173.

18. Fred M. Clough, *The Golden Age, Or the Depth of Time* (Boston: Roxburgh, 1923), p. 40.

19. Edward Bellamy, *Looking Backward, 2000–1887*, ed. John L. Thomas (Cambridge, Mass.: Harvard University Press, 1967), p. 211.

20. Gillette, *Human Drift*, p. 89.

21. Gillette, *World Corporation*, p. 232.

22. Edward Bellamy, *Equality* (New York: Appleton, 1897), p. 54.

23. Howard, *The Milltillionaire*, p. 17.

24. Herman H. Brinsmade, *Utopia Achieved: A Novel of the Future* (New York: Broadway, 1912), p. 39.

25. Gillette, *People's Corporation*, pp. 152, 177.

26. Henry Olerich, *A Cityless and Countryless World; An Outline of Practical Co-operative Individualism* (Holstein, Iowa: Gilmore and Olerich, 1893), p. 74.

27. Thomas, p. 61.

28. Thomas, p. 126.

29. Olerich, p. 350.

30. Byron Brooks, *Earth Revisited* (Boston: Arena, 1893), p. 45.

31. Gillette, *People's Corporation*, p. 161.

32. Edgar Chambless, *Roadtown* (New York: Roadtown, 1910), p. 132.

33. Macnie, p. 114.

34. Gillette, *World Corporation*, p. 134.

35. Thomas, p. 126.

36. Robert H. Thurston, "Progress and Tendency of Mechanical Engineering in the Nineteenth Century," *Popular Science Monthly* 59 (May 1901), p. 38.

37. Kirwan, p. 53.

38. Bellamy, *Looking Backward*, p. 133.

39. Harold Loeb, *Life in a Technocracy: What It Might Be Like* (New York: Viking, 1933), p. 75.

40. Thurston, "Progress and Tendency," p. 37.

41. Robert H. Thurston, "The Mission of Science," *Proceedings, American Association for the Advancement of Science* 33 (September 1884), p. 233.

42. Albert A. Merrill, *The Great Awakening: The Story of the Twenty-Second Century* (Boston: George, 1899), p. 139.

43. Brooks, p. 122.

44. Robert H. Thurston, "Scientific Research: The Art of Revelation and of Prophecy," *Science*, n.s., 16 (September 12, 1902), p. 402.

45. George S. Morison, *The New Epoch as Developed by the Manufacture of Power* (Boston and New York: Houghton Mifflin, 1903), pp. 75–76.

46. Macnie, p. 45.

47. Schindler, p. 111.

48. For too long, however, students of late-nineteenth- and early-twentieth-century America have relied upon *Looking Backward* alone to tell them—and the rest of us—about American culture as a whole at the turn of the century. By their very nature, utopian works deviate from and often distort existing society, not least when their principal purpose is to change it. Consequently, such works do not illuminate the whole of any real-world culture. Rather, they identify particular values, trends, and problems in the culture that fostered them. They must therefore be employed cautiously, as means to full-scale historical inquiries rather than as complete inquiries in themselves. If, then, *Looking Backward* does tell us a good deal about the period in which it was written, as do the other utopian works considered here, it hardly tells us everything we need to know about America then, and neither do those other works.

49. For elaboration, see my "Reconsideration: Harold Loeb's *Life in a Technocracy: What It Might Be Like* (1933)," *New Republic* 175 (October 30, 1976), pp. 42–44.

50. See Warren I. Susman, "The Thirties," in *The Development of an American Culture*, ed. Stanley Coben and Lorman Ratner (Englewood Cliffs, N.J.: Prentice-Hall, 1970), pp. 179–218.

51. On the Technocracy movement, see Henry Elsner, Jr., *The Technocrats: Prophets of Automation* (Syracuse, N.Y.: Syracuse University Press, 1967); William E. Akin, *Technocracy and the American Dream: The Technocrat Movement, 1900–1941* (Berkeley and Los Angeles: University of California Press, 1977).

52. On the World of Tomorrow, see Helen A. Harrison et al., *Dawn of a New Day: The New York World's Fair, 1939–40* (New York: Queens Museum/New York University Press, 1980).

53. On these industrial designers and their contributions to the World of Tomorrow, see Donald J. Bush, *The Streamlined Decade* (New York: Braziller, 1975), chapter 8; Bush, "Futurama: World's Fair as Utopia," *Alternative Futures* 2 (fall 1979), pp. 3–20; Jeffrey L. Meikle, *Twentieth Century Limited: Industrial Design in America, 1925–1939* (Philadelphia: Temple University Press, 1979), chapter 9.

54. On the overall failure of other past technological predictions, see George Wise, Technological Prediction, 1890–1940, Ph.D. diss., Boston University, 1976. Wise's list of "the fifty most frequent predictors" (p. 318) includes Bellamy and Thurston. The predictions by these fifty prophets about technological advances, however seriously flawed, were much more accurate than their predictions about the social effects of those advances.

55. On the rise of "alternative futures," see Alexandra Aldridge, "Imagining Alternative Futures: The Polarities of Contemporary Utopian Thought," *Journal of General Education* 33 (spring 1981), pp. 80–89.

56. On the relationship between technology assessment and environmental-impact assessment, see Harvey Brooks, "Technology Assessment in Retrospect," in *Technology and Change*, ed. John G. Burke and Marshall C. Eakin (San Francisco: Boyd and Fraser, 1979), pp. 465–476. On the relationship between technology assessment and retrospective technology assessment, see Stephen H. Cutcliffe, "Retrospective Technology Assessment: A Review Essay," *Science, Technology, and Society* 18 (June 1980), pp. 7–12.

7

The Home of Tomorrow, 1927–1945

Brian Horrigan

The phrase "home of tomorrow" is highly evocative. The images it calls up today are likely to be either faintly ludicrous or suffused with vague memories of silly cartoons, tourist traps, and pulp science fiction. However, during the period that encompassed the heights of 1920s prosperity, the depths of the Depression, and the patriotic deprivation of the World War II home front, the term "home of tomorrow" was often taken very seriously. At world's fairs and expositions, in department stores, and in home magazines, Americans witnessed an apparently endless parade of predictions about the shape of homes to come.

Whether designs, models, or actual prototypes, these visions of the home of tomorrow had a certain consistency. They represented ideals, and they stood in stark and purposeful contrast to contemporary reality. Their creators were a sizable group of architects, engineers, and businessmen who, acting partly out of a conviction that the housing industry had been mired too long in the bogs of tradition, partly in response to the perceived esthetic dictates of a "machine age," and partly from a desire to stimulate consumption, espoused the idea of the house as a technologically perfected artifact.

During this period, "home of tomorrow" became a kind of code phrase for architects and engineers, a way of identifying their intentions and their broader motivations. A visionary design or model might be used as a device for physically symbolizing a wider vision of the future of housing. Some architects truly believed that they were, in fact, limning the future. For others, the phrase connoted a critique of the present. Professedly futuristic designs could be effective demonstration pieces for new materials or improved building methods. The phrase could also be used in signifying that the work had reached the loftiest heights of au courant modernity.

In the history of the home of tomorrow in this period we can witness architects and reformers struggling to reconcile notions of a design idiom appropriate to the age with tradition-bound sentiment, and we can see their efforts to bring the house—that most recalcitrant of building types—into the mainstream of American technological development. In retrospect, we can discern three scenarios. In one version, architects led or inspired by the European avant-garde would transform the house into a paradigm of modern elegance. In another, engineers or would-be industrialists would clone thousands of cheap dwellings from a single prototype. In the third scenario, the efforts of both the architect and the engineer would be eclipsed by those of the purveyors of consumer goods and gadgets. These scenarios did not follow one upon the other, nor did they remain distinct; they blended with each other and with other aspects of American culture in the interwar years.

The "homes of the future" of these decades cannot be understood apart from a series of interrelated phenomena. First, though housing construction boomed to unprecedented levels in the 1920s, the demand far exceeded the supply, and this frustrating disparity only deepened with the coming of the Depression and the war. The expansive economy and rising standards of living of the early 1920s had heightened the already considerable demands for durable consumer goods. At the head of the list of suddenly indispensable symbols of middle-class status—automobiles, radios, home appliances—was the most durable and elusive good of all, the single-family home. A very real sense of a housing crisis pervaded the era.[1] Second, there was the revolution in design known simply as modernism (or often, in architecture, as the International Style), which had begun in Europe in the wake of World War I. The progenitors of modernism identified their aim as a renunciation of past forms and solutions, and emphatically insisted that design reflect contemporary reality. Humans were racing toward a future of rationality, freedom, and unity with their mechanized, industrial environment, and architecture would provide homes for them. The movement called for far more than structural or formal revolution. Always implicitly and sometimes explicitly, modernist architects demanded a total reconsideration of the form and function of the dwelling. "The house is a machine for living," announced the Swiss architect Le Corbusier, providing the movement with its most notorious slogan and the clearest statement of its revolutionary goals. With hindsight, one can recognize in the late 1920s a zenith in the arc of modernist design and thought in Europe. At the same time, the style and its accompanying rhetoric were attempting to take root in American soil. In the United States, modernism was identified literally as a visionary style. The shocking designs that modern architects boldly cast into the American landscape were widely apprehended as true rehearsals for the future.[2] Finally, the "housing futures" offered during the period were ineluctably linked with machines and mass production, though each option represented a different interpretation of this relationship. The spectacular success of the techniques of mass production (which had been largely responsible for the prosperity of the 1920s) began to tempt engineers and architects to transfer these techniques wholesale to the housing industry. Many observers believed—or feared—that the market for new cars was reaching a plateau, and some proposed replacing it with a market for housing manufactured on an industrial scale; like automobiles, houses would roll off assembly lines by the thousands in affordable, accessible packages.[3] Architecture critic Theodore Morrison wrote in *House Beautiful* in 1929: "Until our houses can be made in the factory, by machine, we shall have no true economy of housing comparable with the economy prevailing throughout industry generally. Until they can be installed, not built, we cannot expect them to be truly efficient and rational adaptations of means to an end."[4] What had charged Morrison with this conviction was the recent appearance of the most radical dwelling machine of the day, R. Buckminster Fuller's Dymaxion House. Fuller's brilliant, quixotic design so thoroughly established a pattern for all subsequent futuristic speculation on the home that it is worth examining in some detail.

In 1927, Fuller was a young, unemployed, mostly self-taught engineer bursting with revolutionary notions about the present and the future of the American housing industry. That year, Fuller published his views along with a detailed description of a house for mass production in a 50,000-word broadside entitled *4-D*.[5] He sent copies to friends and family members and to such luminaries as Henry Ford, Bertrand Russell, and advertising man Bruce Barton. To dramatize his ideas, Fuller constructed a large model of his proposed house, which he called a 4-D Utility Unit.

On the reasoning that the home of the future should be lightweight and easily demountable, Fuller's design featured a central aluminum "mast," from which transparent glass and casein walls and inflated rubber flooring were to be suspended by wires. The mast (the analogy to naval architecture was intentional) was to contain all the household services—one of the first instances in domestic planning of the "service core." In the core were to be two bathrooms (complete with vacuum electric hair clipper, vacuum toothbrush, and chinning bar), a self-activating laundry unit that would deliver washed and dried clothes

Figure 1
The Dymaxion House (4-D Dwelling Unit) of 1927 marked Buckminster Fuller's sensational debut as a visionary engineer and architect. Buckminster Fuller Foundation, Los Angeles.

in 3 minutes, sewage disposal tanks, an electric generator, an air compressor, a humidifier, and a kitchen with every conceivable appliance. The two small bedrooms were to have pneumatic beds with neither sheets nor blankets, these being unnecessary in the perfectly climate-controlled house. With drudgery eliminated, wrote Fuller, "the real individualism of man and his family may be developed . . . creation will set in as never before." To this end, space was set aside for a "creative room" (or a "get-on-with-life" room, as Fuller also called it), which would be equipped with a typewriter, a calculator, a telephone, a dictation machine, a television, a radio, a phonograph, and a mimeograph machine, all in one factory-assembled unit. In the space beneath the house would be stored the family "transport unit," an amphibious auto-airplane.[6] The houses were to be capable of being stacked on a single tall mast to make up an apartment tower. Singly or in stacks, they could be moved, preferably by zeppelin, and plugged in anywhere. Since each house would have its own power generator and a recycling water system, attachment to public utilities would be unnecessary. In fact, housing in the Dymaxion Age would be very much like a utility. Produced by the millions, houses would be provided as public utility companies provide gas, electricity, and telephones. Obsolete equipment would be replaced by improved models.

Today, it seems almost preordained that Fuller's public technological spectacle should have appeared in 1927. Lindbergh's flight, the first talking movie, the establishment of transatlantic radio-telephone service, the first public demonstration of television, the opening of the Holland Tunnel, and the dramatic appearance of Henry Ford's Model A all occurred in that year, producing a kind of mass hysteria.[7] Bucky Fuller, living in raucous, industrial Chicago, no doubt felt the day's fervid excitement for science, machinery, and massive industry.

Fuller's model of the Dymaxion House was first shown to the public in April 1929 at Marshall Field's department store in Chicago. It was used as the centerpiece of a display of modern furniture from the 1925 Paris Exposition des Arts Decoratifs, from which had emanated the earliest waves of modern design to break on American shores. The display, accompanied by lectures delivered by the voluble Fuller six times a day for two weeks, caused a sensation.

The Dymaxion House was a bolt out of the blue. Amid all the futuristic clangor of technological newness in the late 1920s, there had as yet been no murmur about that sacrosanct American institution, the home. Fuller changed all that. Here, in 1929, was a shocking vision of the future of the home. Shocking, yes; but, given all the recent invasions of technology into private life, perhaps believable.

The Dymaxion House gained instant notoriety for its creator. Everywhere Fuller appeared with his model he attracted hundreds of curiosity seekers and a great deal of media attention. Fox Movietone released a newsreel featuring the model, and scores of newspaper and magazine articles heralded the Fuller future: "Foresees Home Made in Factory—Ready to Occupy;" "The House of the Future;" "House in Utopia;" "Machine-made Family Life;" "Modern Houses will be Built for $3,000;" "Home in the 21st Century;" "Homes You Will Carry With You When You Move;" "Bedsheets Unnecessary in House of the Future;" "House of 1982 Built Like Ship;" "Everyman's House."[8] Since Fuller's model was usually exhibited in a gallery, in an artist's studio, or at an arts club meeting, many saw the house as the latest event in a continuing artistic revolution; "the most exciting art idea in centuries" a Chicago paper called it.[9] Others concentrated on the consumerist implication of a house full of gadgetry. Still others saw the house as a manifestation of the "world-wide '100% sunshine and fresh air' movement," of the "nudism and sun-bathing cults that have sprung up on all five continents"[10]—a notion Fuller had encouraged by placing a few naked dolls in the model.

An artistic watershed; a technological paradise; a luminous, healthy, liberating environment—these were all promised by the Dymaxion House, and constituted something of a thematic checklist for subsequent homes of the future. But Fuller's intent was deeper still. As much as it was a prediction of the future, the Dymaxion House was a polemical assault on the present. Fuller's model was an effective bit of agitprop, which he used to inveigh against a housing industry that was complacent and ineffectual.

Fuller fervently believed in the essential moral rightness of his crusade, in the supposed honesty of industrially designed objects. The transcendentalism and technological determinism of *4-D* place its author firmly in the American grain:

> As mechanical truths are revealed, so do we progress toward perfection: though there can be no absolute perfection in the material world. So has the automobile or airplane continually approached perfection. As it has approached perfection by the process of the application of truth, so has it approached one final design. . . . Just so is there a final best design, in our mechanical age, of the home or living quarters. Eventually, through economic pressure, and the desire of mankind for individual abstract expression, property ownership and travel, this home will come.[11]

Mass production was for Fuller a kind of redemptive force, the best path to true rationalization of the housing industry. Of course, standardization had played an important role in American domestic architecture since the middle of the nineteenth century. In the 1840s, prefabricated (or "pre-cut") wooden balloon-frame houses were being shipped to all parts of the country. By the 1920s, mass-produced ornament, doors, windows, stairs, and household equipment had become indispensable elements of the average American house.[12] To Fuller, and to virtually every other critic of the period, standardization applied in this fashion had degraded the American house. However, it was thought that standardization, reinterpreted and controlled, could reverse this decline and elevate the house to a state of modern perfection.

The possibility of transforming the housing industry through mass production captured the imaginations of many architects, engineers, industrialists, and social commentators. In 1932, *Fortune* magazine published a series (unsigned, but written by Archibald MacLeish) on the disastrous state of American housing. In the final article, entitled "Solutions," MacLeish wrote: ". . . standardization by factory production with expert engineering and complete functional efficiency is certainly preferable to the standardization by sheer imitation and inertia which is as visible today in the mock European houses of the fashionable suburbs as in the mock suburban villas of the subdivision lots."[13] Thoroughly convinced by the scenario Fuller had outlined, MacLeish asserted: "It is now past argument that the low-cost house of the future will be manufactured in whole, or in parts, in central factories, and assembled on the site. In other words, it will be produced in something the same way as the automobile."

In basing their rhetorical model on the automobile assembly line, the paragon of American industrial success, the proponents of efficient housing production were helping to create what David Hounshell has termed an "ethos of mass production." A clear expression of this ethos was already available by 1925 in a lengthy treatise, *The Way Out: A Forecast of Coming Change in American Business and Industry*, in which successful Boston businessman and self-styled prophet Edward A. Filene argued for the universal application of Ford's methods, claiming: ". . . a Fordized America built upon mass production and mass distribution will give us a finer and fairer future than most of us have dared to dream."[14] Thus, the slogan "Houses Like Fords," which became popular at the end of the 1920s and in the early 1930s, was more than just a catchy headline; it was a rallying cry and an agenda for the future.

In the development of an ideology and a tangible imagery of mass-produced homes of the future, the contributions made by European architects of the "modern movement" were critical. In 1923 Le Corbusier issued this exhortation in his tract *Vers une architecture*: "Industry on the grand scale must occupy itself with building, and establish the elements of the house on a mass-production basis."[15] The English translation of Le Corbusier's manifesto, published in 1927, had an enormous impact on American architectural consciousness, but no exemplar of the movement's goals was built in the United States until 1929, when architect Richard Neutra completed his early masterwork, the Lovell House in Los Angeles. Combining industrial imagery and construction methods (welded steel frame, flat roof, ribbon windows, and sprayed concrete walls), the Lovell House remains the archetypal modernist house.

Neutra, who had emigrated from Germany in 1923, found in Dr. Philip Lovell the ideal client for his unstintingly avant-garde architecture. A well-to-do, free-thinking "naturopath," Lovell, as Neutra later wrote, "wanted to be a patron of forward-looking experiment.

Figure 2
Richard Neutra's elegant, severe Lovell House, completed in 1929, remains the archetypal "machine for living" in the United States. Richard and Dion Neutra, Architects, Los Angeles.

He would be the man who could see 'health and future' in a strange wide open filigree steel frame, set deftly and precisely by cranes and booms into this inclined piece of rugged nature."[16] Neutra and his fellow modernist émigrés had expected to find such clients in the United States. Americans, they thought, would be more receptive than Europeans to the architecture of the machine age because they had so easily integrated the principles of mass production into their lives. A later student of Neutra, California architect Harwell Harris, said: "It would not be far-fetched to think that Neutra came to America because America was the home of Henry Ford. Ford was more amazing to Europeans than to us, who saw in him our own features."[17] Nevertheless, for all its unconventional characteristics, the Lovell House was anything but an industrial prototype. On the contrary, it was an expensive, hand-crafted, machine-age mansion. It was the *image* of mass production that Neutra invoked, down to the Model T headlights used as interior lighting in playful homage to Henry Ford.

Notoriety instantly accrued to the houses built by Neutra and other modernists. Though their number was relatively small, their stunning, iconoclastic appearance assured them a place in the public eye and convinced many that these were indeed the advance guard of a whole new race of buildings. Though critics caviled that the "so-called International Style house developed in America" was "merely the old house at the old price in a new envelope," these houses were emotionally, if not intellectually, persuasive as clairvoyant designs for living.[18]

By 1930, then, a dual image of the home of tomorrow had been developed before an expectant public. On the one hand was the luxurious "machine for living" of the modern movement, on the other the cheap, identical, machine-made house of Fuller and other proponents of mass production. Both had great potential appeal in Depression America, the one offering escape into a voluptuous Hollywood future and the other promising industrial recovery and universal homeownership. Though they shared an ideological foundation in American technological utopianism, the two images existed in uneasy tension.

This tension emerged very clearly at the Century of Progress Exposition in Chicago in 1933 and 1934. Innovations in housing had been a part of the displays at world's fairs and expositions since the World Exhibition at London in 1851, but at no fair before the Century of Progress Exposition had housing (specifically, the single-family house) played such an important role. Some observers expected that the fair would have as profound an effect on the course of American housing as the Columbian Exposition of 1893 had had on American architecture in general. Five acres of the lakefront site were given over to a section called "Home and Industrial Arts." Thirteen full-scale, furnished model homes were exhibited, nine of which were intended to represent some approach to prefabrication. In fact, it was with this heavily publicized exhibition that the term *prefabrication* came into general use as a substitute in discussions of housing for *mass production*.[19]

The star attraction at the Century of Progress Exposition was George Fred Keck's "House of Tomorrow." Built around a central utility core, the twelve-sided, steel-framed, completely glazed structure strongly recalled the Dymaxion House.[20] Keck's house also looked "industrial," though it was built on the more conventional principle of compression rather than on Fuller's suspension system. On the ground floor were a recreation and work room, a garage, and a hangar for the family airplane. Modern, custom-designed furnishings in rich materials (leather, walnut, ebony, mahogany, and chrome) abounded. The House of Tomorrow, which cost 10 cents to see, was financially the most

Figure 3
The "House of Tomorrow," designed by George Fred Keck, was erected for the Century of Progress Exposition in Chicago in 1933. Hedrich-Blessing.

successful house at the fair. For most of the 750,000 people who saw this dazzling structure, it was the first real encounter with that "machine for living" they had been hearing about.

Though Keck's House of Tomorrow was certainly a radical design, it remains difficult to imagine it in serial production, at least at a price affordable by any but the wealthy. Keck's image of the future, at least in this instance, was one of luxury. He seemed to be following the lead of modernists such as William Lescaze, a Swiss architect who, like Neutra, had worked in the United States since 1923. In 1928 Lescaze had published a design for an extravagant "Future American Country House" in the machine-for-living idiom.[21] These architects invoked the future not to signal a revolutionary program of mass-produced houses but rather to call attention to the position of their work on the cutting edge of design and to appeal to a fashion-conscious clientele.

Keck built another house—the "Crystal House"—for the second year of the Chicago fair with his share of the profits from the House of Tomorrow admission fees. By 1934, he had felt the need to make some gesture toward producing a house that "lends itself to prefabrication," as he explained, "in order that it may be within reach of the masses." Keck estimated that, with a production run of 10,000 units, this house could be constructed for $3,500 or less.[22] The Crystal House was essentially a glass box suspended within a structural steel cage. Completely air-conditioned, it was filled with appliances and expensive copies of Bauhaus furniture. As in the House of Tomorrow,

Figure 4
William Lescaze's 1937 design for the "House of 2089" was a streamlined reworking of an earlier futuristic project, and included a "auto-giro" parked on the flat roof. Avery Library, Columbia University.

Figure 5
George Fred Keck's entry for the second year of the Chicago fair was this "Crystal House." Iconographic Collections, State Historical Society of Wisconsin.

Figure 6
General Houses, Inc., founded in 1932 by architect Howard Fisher, promised to mass produce houses by adopting the techniques of the automobile industry. This full-scale prototype was exhibited at the Chicago fair in 1933. Hedrich-Blessing.

rich woods and chrome were used in the interiors. Suggestively displayed with the house was Buckminster Fuller's three-wheeled, teardrop-shaped Dymaxion Car. Though the innovative construction techniques, rapid erection, and industrial materials of the Crystal House undeniably implied mass production, the house as exhibited—and thus as publicly perceived—was hardly more reasonable as a prototype of mass-produced things to come than its predecessor. As critics pointed out, few of the promises of prefabrication made at Chicago were kept. Such isolated reveries as the Crystal House provided no realistic responses to the universally acknowledged shortage of affordable dwellings.[23]

The house at the Century of Progress Exposition that seemed to housing observers to be the true augury of the future was a small, self-effacing entry from a new company, General Houses, Inc., the brainchild of 30-year-old architect Howard Fisher. This one-story, single-family house was meant for reproduction on a vast scale. With its flat roof, steel casement windows, and enameled steel panel walls, the unornamented house was the epitome of the trim little machine. To Fisher, the frankly modern aspect was imperative for both structural and stylistic reasons. At a symposium on prefabrication sponsored by *House and Garden* magazine in 1935, he said: "The public's conception of style has heretofore been much more advanced in the case of their automobiles, for example, than in the case of their houses; but I believe they are going to develop an equally advanced design in houses. . . . I believe the greatest selling point these houses will have will be style."[24]

General Houses initially offered eighteen variations on the basic model exhibited at the fair, each coded with a "chemical formula," such as "K_2H_4O," which identified the particular type and gave it the cachet of scientific research. The small models were to sell for about $4,000, and General Houses anticipated expanding its repertoire to include larger houses—the Cadillacs or Lincolns of the line.

General Houses was one of a handful of prefabricated-housing organizations to enjoy a small measure of success during the Depression. Most architects and engineers who applied themselves to the problem of the house during this period, whether concentrating on new design or innovative housing equipment, shared a significant shortsightedness: By focusing on the house itself, they failed to respond to the wider range of public demands and expectations. To most prospective home buyers, problems in design and construction paled in comparison to the problem of getting a home at all. It was "not so much a lack of raw materials or skill that creates the 'social problem' of a housing shortage," wrote Robert and Helen Lynd in *Middletown*, their 1929 study of a Midwestern community, as the "the intricate network of institutional devices through which a citizen of Middletown must pick his way in undertaking the building of houses for others or in trying to secure a home for his own use."[25]

Some housing prophets of the early 1930s took the automotive-marketing model to extremes, forecasting a time in the "not-far-distant-tomorrow" when

> All the home owner need do is visit the "Home Headquarters and Architects Clinic" in his own hometown or in a nearby city. There, houses that are within his price range will be presented for his inspection. He will concentrate on the planning and will consummate the purchase by signing documents. But these documents need be no more complicated than that signed by today's purchaser of a car on the installment payment plan.[26]

In planning General Houses, Howard Fisher had perceived these problems. What distinguished General Houses was not its hardware but its broader vision of the housing problem. As an architecture student at Harvard in 1929, Fisher had been inspired by Fuller's ideas on the future of housing as a service industry. The youngest member of a prominent family from Chicago's North Shore (his father, Walter L. Fisher, had been Secretary of the Interior under Taft), Fisher saw with a clarity and an honesty unusual among 1930s prefabricators that the housing problem was "many separate problems which affect and are affected by the current economic, social, and political order."[27] By the spring of 1932, General Houses had been organized as the "G.M. of the new industry of shelter," with architects, engineers, manufacturing executives, realtors, and lawyers on its staff. It proposed to market an entire housing business in all its phases: research, advertising, legal counsel, financing, land control, landscape architecture, community planning, and interior decoration, as well as the development of prefabricated houses.[28] General Houses intended to act, as did many automobile manufacturers of the day, as an assembler of parts rather than a primary producer. The company enlisted the participation of an impressive battalion of manufacturers, including the Pullman Company, the Container Corporation of America, Inland Steel, Curtis Companies, Inc., American Radiator, Pittsburgh Plate Glass, General Electric, and Thomas A. Edison, Inc.

For a time after the Chicago fair, a "prefabrication bandwagon" rolled through American business.[29] The mood was feverish, at times carnivalesque. A new prefabrication company, American Homes, Inc., unveiled its "Motohome" with considerable fanfare at Wanamaker's department store in New York on April 1, 1935, with Sara Delano Roosevelt, the president's mother, cutting the ribbon on a house wrapped in cellophane, that newest and most futuristic of materials.[30] The Motohome was named for its so-called Moto-unit, a central service core between the kitchen and the bathroom containing the plumbing, heating, and electrical equipment. The complete but unassembled house would arrive at its site in a truck labeled "This Truck Contains One American Home." "The packaged home is here," proclaimed Katharine Bissell, a writer for *Women's Home Companion*, "and will soon be exhibited at your favorite department store. Imagine being able to buy your home as you would buy a package of cereal or face powder—a home complete in every minute detail, that you can actually see, touch, examine, and discuss before you buy it, and above all know exactly what it is going to cost, down to the last penny before you move in."[31] American Homes took the concept of packaging to new heights. Each house came equipped with a two-day supply of groceries in the kitchen and a complete set of home advice manuals, with tips for the homemaker on home maintenance, decoration, landscaping, household budgeting, cooking, etiquette, and child-rearing.[32]

In short, one bought not only a house but an entire code of middle-class respectability.

The Motohome's architect, Robert McLaughlin, had begun his career designing lavish country villas for wealthy clients. However, by the late 1920s he had, as *Fortune* was later to phrase it, "heard the cries for low-cost housing, and conceived the possibility of a tremendous market and an architectural practice far more exciting than the confection of upper-crust manor houses—and far more lucrative as well."[33] McLaughlin formed a partnership with housing advocate Arthur C. Holden and began to study the problem of low-cost housing. In 1933, after the success of twenty experimental prefabricated houses for coal workers in Pennsylvania, McLaughlin founded American Houses, Inc. The next year, American Houses joined forces with Houses, Inc., a company founded with the backing of the chairman of the board of General Electric, Owen D. Young, and led by self-taught engineer and entrepreneur Foster Gunnison. Houses, Inc., built no houses of its own but rather was set up as a holding company to acquire and promote independent prefabrication systems. Gunnison orchestrated the spirited publicity campaigns for the Motohome and was the source of most of the various sales gimmicks, including the food in the kitchen, the advice manuals, and the toilet seat that "weighs the sitter when he raises his feet."[34]

Fisher, McLaughlin, and Gunnison typified the youthful architect-entrepreneurs who broached the possibility of a future of prefabricated homes. However, several established architects turned their sights in this direction in the 1930s. Richard Neutra, already famed as the preeminent designer for the wealthy avant-garde in Southern California, introduced his designs for a "Diatom One + Two" house based on a Fulleresque suspension principle and having walls of a unique diatomaceous-earth composition.[35] William Van Alen, the flamboyant architect of the Chrysler Building, designed a severe "steel-shell" house for a new company, National Homes, Inc., in 1935, and erected it, incongruously, on the corner of 39th Street and Park Avenue in Manhattan.[36] Frank Lloyd Wright, his career in eclipse since the early 1920s, began building his "Usonian Houses" in 1934. Though still designed as individual commissions, the small and affordable Usonian Houses were also meant by Wright as exemplary models of the homes that would one day multiply across the landscape of his epic, futuristic "Broadacre City."[37]

Most of this movement was smoke, with very little fire. Broadacre City remained a compelling paper vision. The little house on Park Avenue disappeared, taking National Homes with it. Neutra's Diatom house was never built. The ambitious program that Howard Fisher had sketched in 1932 never materialized. None of the more enlightened features of General Houses—the financing, the legal services, the community planning—were ever instituted, and the company became a supplier of conventionally styled though still largely prefabricated houses. Though American Homes survived well into the 1940s, fewer than 150 Motohomes were ever actually built; its later products, like those of General Houses, grew more conservative in shape and less idealistic in content. In 1936 the National Association of Housing Officials reported that, in spite of a "general impression" that prefabricated houses had solved the problem of better building at lower costs, a thorough survey of the market revealed "no immediate prospects for mass production."[38] In the last five years of the 1930s, when housing starts were slowly climbing again, fewer than 1 percent of all new single-family houses were prefabricated.[39]

Thus, the belief that the future of the house could be enacted through mass production faded in the same decade in which it had so confidently arisen. Perhaps the most serious failing of this vision was that it focused on a single, traditional building type—the free-standing single-family house. It was as true in the 1930s as it is now that housing must be investigated not solely as a technological issue but as one that is intimately woven into the social fabric. In 1931, Lewis Mumford chided proponents of the factory-made home for promoting an "unworkable anachronism," which he archly defined as "a communism of technique in the production of houses, with the usual anarchy and monopoly in our system of land holding, financing, and community design." Mumford persuasively argued that "no decent 'house of the future' can be designed in the factory alone. To forget this is to foster specious hopes. . . ."[41] Mumford was at the center of a group of architects and social critics who had been arguing for a new "social economy of housing" since before World War I. The efforts of this group of "housers" (as they were sometimes called) to promote new ideals of community planning eventually bore fruit in the New Deal housing programs, especially the Greenbelt Town Program. The effect of these programs on the public consciousness may have helped deflate the dreams of the prefabrication entrepreneurs. Housing and community planning were moving (or so it seemed in the mid 1930s) into the realm of government control and public policy. The future began to look very different.[41]

A second reason for the failure of the dream of mass-produced homes lay in the simple fact of inadequate capital. Buckminster Fuller realized that to tool up for mass production on the scale necessary to achieve truly affordable houses would require an immense outlay of capital. When officials of the Century of Progress Exposition approached Fuller with the proposal that he exhibit a full-scale mockup of the Dymaxion House, he refused, saying he would be satisfied only if a true prototype for mass production could be built, the cost of which he estimated at $100 million.[42] Enthusiastic tyros such as Howard Fisher and Robert McLaughlin may have subconsciously identified with the myth of the heroic, youthful innovator creating a vast, new, and necessary industry—a myth that Henry Ford had created almost singlehanded. But American business policy had become so thoroughly governed by the tenets of corporate capitalism by the end of the 1920s (if not much earlier) that to have anticipated success for a sweeping revolution in such a critical industry without massive assistance from major corporate sponsors was self-delusion at best.

A third and related reason for the failure of the vision lay in the fact that many large companies were indifferent or even hostile toward it. As already mentioned, General Electric offered brief initial support to the production of the Motohome, in a rather tenuous alliance made on the basis of an old friendship between G.E. chairman Owen Young and Foster Gunnison. However, dissent among the principals of American Homes led both Gunnison and General Electric to sever their ties with the company. Likewise, none of the impressive corporate associates of General Houses ever acted as true sponsors or underwriters; their commitment had been only to supply parts. Among steel companies there had been a flurry of interest in the possibilities of mass production of steel houses in the early 1930s, and several major concerns had built experimental houses, such as American Rolling Mills' "Ferro-Enamel House" at the Chicago fair. But the future of steel houses, which many had predicted in glowing terms, also failed to arrive.[43] Another large corporation, American Radiator Company, was rumored in the early 1930s to have a major prefabricated housing program waiting in the wings. Yet corporate timorousness about the field's possibilities was such that the American Radiator test house was "shrouded in secrecy" on the roof of the company's skyscraper headquarters, and the plan was apparently shelved.[44]

Corporate reluctance matched public sentiment and taste. Even in a hypothetically clear field, it is doubtful that American corporate interests would have backed the development of a mass-produced house, so antithetical did it seem to traditional patterns of American domestic life. Americans may have flocked by the thousands to the department-store and exposition displays, but they went to look, not to buy. Their initial curiosity about the machine style never ripened into complete acceptance, Howard Fisher's sanguine predictions notwithstanding. Even after the prefabricators capitulated to the public preference for peaked-roof wooden houses, the products stayed in their crates.

Although, like the public at large, American corporations ultimately refused to underwrite a future full of modernist mansions or mass-produced homes, they were attracted by a shinier side of the Home of Tomorrow coin: the house as a wonderland of gadgets. It is not surprising that the companies that associated themselves most readily with the Home of Tomorrow were the major manufacturers of electrical appliances. General Electric exhibited a "House of Magic" at most of the major fairs of the 1930s.[46] Alleged to "walk and talk," the house was not really a separate structure but a gimmicky update on the department store "demo" home, a kind of stage set on which glamorous women were cast as housewives, running the household machinery and making a sales pitch. Westinghouse, not to be outdone, built an entire "Home of Tomorrow" in 1934 in Mansfield, Ohio. It was intended as a lived-in laboratory in which the company's engineers and their families would temporarily reside to test the equipment. This house, a tour de force of household electrification, was designed to attract attention, which it did quite effectively. Designed by architect Dwight James Baum, the house was a conventional wood-frame and stucco structure, only slightly odd in style—a sort of Regency-Cubist affair, employing virtually none of the already notoriously "futuristic" modern vocabulary. Indeed, architecture was quite beside the point, according to the Westinghouse engineer responsible for the house. "A new profession of 'house engineers,' " maintained Victor G. Vaughan, "will soon absorb all architectural functions except those of a purely aesthetic nature."[46] The engineers had a field day with the Westinghouse prototype, providing a connected electric load equal to that of 30 average houses, "ready to do the work of 864 servants with the flip of a switch." Some of the features of the house were air conditioning, an electric garage-door opener, automatic sliding doors, an electric laundry, 21 separate kitchen appliances, burglar alarms, 140 electrical outlets, and 320 lights. All this was available, or so it was claimed, for around $12,000. Westinghouse admitted that the price would probably place the house beyond the means of most families in the future, thus further removing this spectacular exercise from the democratic rhetoric of the prefabricators.[47]

A consumerist orientation was evident to some extent in every Home of Tomorrow during this period, even those with more revo-

Figure 7
The Westinghouse Corporation built this "Home of Tomorrow" in 1935 as a showpiece for futuristic electronic gadgetry for the household. Westinghouse Historical Collections.

lutionary intent. It should be recalled that Fuller's Dymaxion House—that altruistic vision of Everyman's dwelling—made its debut at the very crossroads of Chicago consumer society: Marshall Field's department store. At the Century of Progress Exposition, most of the demonstration houses were advertising showcases for producers of building materials, decorating firms, or department stores. Such commercialism tended to transform the houses into containers of desirable consumer items, diverting attention from them as radical experiments in design or industrial process.

This dominance of consumerism hints at a final reason for the failure of the mass-produced home to fulfill its promise. By the end of the 1930s, a sharp split had occurred between the popular perceptions fostered by avant-garde architects and their glamorous "homes of the future" and the nagging reality of the housing situation. The most thrilling homes of the future were precisely those whose future realizations would forever elude the grasp of most people. Neutra's Hollywood mansions and Keck's stunning exposition ventures were suggestive and seductive dream houses, wholly unreasonable as models for replication. The public, filled with great expectations, was un-

Figure 8
Among the more highly touted marvels of the Westinghouse "Home of Tomorrow" was the electric garage-door opener. Westinghouse Historical Collections.

doubtedly disappointed with the choices that actually became available, though these choices bore the same "home of tomorrow" label. The typical product of General Houses or American Houses looked cramped and miserly next to the suave designs of George Keck or the electrified fantasy of the Westinghouse engineers. If the public was expected to make the bold leap into the future, to become proselytes of the modernist faith, it had to be offered more than the lowest common denominator that the minimal "prefabs" seemed to embody.

Insofar as the phrase "home of tomorrow" implied a promise of personal attainment, it had a hollow ring for most Americans during the Depression. The prediction of revolution in industrial production was also a gross miscalculation. The phrase "______ of tomorrow" gradually lost its connotations of prediction, prescription, and solution and was indiscriminately applied to everything from new floor coverings to vacuum cleaners to "revolutionary" plumbing fixtures. By the end of the 1930s, the phrase was merely an advertising slogan, designed to stimulate buying, to condition consumers to accelerated rates of change, and to promote expectations of newness.

It was also clear by the late 1930s that future houses would not necessarily be "modern" or avant-garde in design. The Modern style was not wholly rejected; rather, it became one of several available "period styles." Like Colonial, Tudor, and Spanish houses, the Modern house wore a romantic costume, but one based not on the past but on a fantasy future. What failed to thrive in the American consumer atmosphere was not the image that the movement's leaders had fashioned but the deterministic and often alarmingly revolutionary rhetoric that had originally underpinned it.

Assertions that modernism was inevitable because of its technological and moral perfection were most frequently heard in the early years of modernism's influence in the United States, when the discourse was dominated by the more doctrinaire Europeans. By the middle of the 1930s, attempts were being made to reconcile the alien doctrines with the American character, to justify them in terms of consumerism. For example, in 1935 *Fortune* magazine (an early and loyal champion of modern architecture) praised the modern house as "the house that works," confidently predicting for modernism a bright commercial future.[48] The editors asserted that the marriage of modern form and modern (that is, highly mechanized) content would be happy and inevitable, declaring: "Modernism in America will be full of gadgets because modernism in America will be gadget's child." In other words, Americans' well-known love of household gadgetry would be admirably requited by the mechanistic modern style; gadgeteering had at last found its most appropriate arena. But *Fortune*'s tautology missed the mark. The modern style did not bring with it, perforce, an equally advanced battery of household equipment, nor was the inverse necessarily true. It became clear that modernity in houses had nothing to do with avant-garde architectural style or mass production; the modern house was simply the well-equipped house. For the most part, Americans did not want machines to live in; they wanted machines to live with.

This proclivity was quickly confirmed. During World War II the most popular "home of the future" was not a house at all but only what we might call a high-tech kitchen. Three full-size mockups of the Libbey-Owens-Ford company's "Day After Tomorrow's Kitchen" circulated to department stores all over the country in 1943 and 1944. Over 1.6 million spectators "beheld what the future had in store," in

Figure 9
The "Kitchen of Tomorrow" of the Libbey-Owens-Ford Company, circulated to department stores in 1944 and 1945, promised housewives a near future of ease and luxury. Hedrich-Blessing.

the words of historian Siegfried Giedion, who probably beheld it in person while writing his seminal *Mechanization Takes Command*.[49] With its sleek surfaces, glass-walled refrigerator, dishwasher, sunken saucepans, and recessed waffle iron, the L-O-F kitchen was a tantalizing display, conditioning an eager body of consumers for the glittering prizes that awaited them at war's end.

Designers and architects clearly were rethinking their commitment to modernism and the future of the American home, and many were repenting for their earlier futuristic excesses. Walter Dorwin Teague, one of the key figures in the revolution in industrial design of the 1930s, wrote an article in 1943 for *House Beautiful* called "Sane Predictions about the Houses You Will Live in after the War," the clear implication being that earlier predictions about the near future of the home had been irrational.[50] In the same year, Kenneth Stowell, the influential editor of *Architectural Record*, wrote a provocative editorial entitled "The House of the Future, 194?–195?,"[51] which reads like a keynote address to a convention of chastened architects. Some of his aphoristic prophecies:

The house of the future will perform the same functions as the house of the past and the house of the present. . . .

The house of the future will have floors, walls, ceilings, partitions, and roofs. . . .

In appearance, the houses will reflect the desires, tastes, associations, prides and prejudices of their owners. . . .

Radical experiments and designs will continue to intrigue those who want to be in the vanguard of progress. Conservative designs, reflecting the best of the past, will be built to please those who prefer the familiar. . . .

The house of the future will still be a house.

If American architects and homebuyers in the 1940s adhered to familiar images, it was understandable. In those emotionally shattering years, styles that evoked a reassuring past seemed ever more important, even for houses of tomorrow. And the conservative urge continued when the war ended. When Mr. Blandings builds his dream house in the comic novel and movie of 1946,[52] he rejects his architect's flashy modern suggestions and opts for a classic New England Colonial design. The episode is meant to be funny, but not surprising. Mr. Blandings (the name is a giveaway) knows what he likes: a house that is tasteful, comforting, and just like everyone else's.

In later years, new shapes of things to come emerged, each accompanied by promises of appropriateness and universality. From the curvaceous molded plastic structures of the 1950s, through the urban homesteads and rural geodesic retreats of the 1960s, through the passive-solar "cluster" homes of the energy-conscious 1970s, and on to the electronic cottages of the 1980s, "homes of tomorrow"—bright and hopeful packages of new materials, technologies, enthusiasms, and anxieties—have appeared regularly. Everyone seems oblivious to the historically low success ratio of such predictions. The "home of tomorrow" appears to be a fixture of American capitalist society, but it seems destined to be always just over the horizon.

Acknowledgments

For research support, I am grateful to the Mabelle McLeod Lewis Foundation of Stanford, California, and the Smithsonian Institution, where I was a predoctoral fellow in 1980 and 1981. In a much reduced version, this paper was presented at the annual meeting of the Society for the History of Technology in Philadelphia in 1982. I am grateful to Joe Corn for his careful reading and editing of the many versions of this paper. I am also indebted to Kathleen Horrigan and Amy Levine for their help.

Notes

1. Gwendolyn Wright, *Building the Dream: A Social History of Housing in America* (New York: Pantheon, 1981), pp. 193–214; John Gries and James T. Ford (eds.), *Publications of the President's Conference on Home Building and Home Ownership*, 11 vols. (Washington, D.C., 1932).

2. Henry-Russell Hitchcock and Philip Johnson, *The International Style* (New York: Norton, 1966; originally published as *The International Style: Architecture Since 1922*, New York, 1932); Reyner Banham, *Theory and Design in the First Machine Age*, second edition (New York: Praeger, 1960); William Jordy, "The Symbolic Essence of Modern European Architecture of the Twenties and its Continuing Influence," *Journal of the Society of Architectural Historians* 23 (1963), pp. 177–187.

3. David Hounshell, *From the American System to Mass Production, 1800–1932* (Baltimore: Johns Hopkins University Press, 1984), pp. 303–330.

4. Theodore Morrison, "House of the Future," *House Beautiful* 66 (September 1929), p. 292.

5. R. Buckminster Fuller, *4-D Time Lock* (Albuquerque: Lama Foundation; originally published privately as *4-D*, Chicago, 1927). The exact chronology of the various versions of the house and the manifesto is not clear from the sources. Fuller always held that 1927 was the date of the project's creation, though he did not apply for a patent until 1928, which was also the year he copyrighted *4-D*. "4-D" refers to time, the "fourth dimension," and was Fuller's first name for the house. "Dymaxion" was a word coined by publicists for Marshall Field's department store, where the model was first exhibited publicly. The P.R. men considered the term—combining "dynamism," "maximum," and "ions"—catchier and more scientific-sounding. Fuller, fond of neologisms, obviously liked the word, and it became a kind of trademark for his entire career.

6. Fuller, *4-D*, pp. 21–23.

7. See Robert A. M. Stern, *George Howe: Towards a Modern American Architecture* (New Haven: Yale University Press, 1975), pp. 71–78; R. A. M. Stern, "The Relevance of the Decade," *Journal of the Society of Architectural Historians* 24 (1965), pp. 6–10; Allen Churchill, *The Year the World Went Mad* (New York: Crowell, 1960).

8. In the order listed, the articles appeared in the following newspapers and journals: *Chicago Tribune*, May 7, 1929; *New Republic*, May 31, 1931; *Time*, August 22, 1932; *The Architect* July 1928; *Milwaukee Sentinel*, May 27, 1932; *New York Daily News*, October 9, 1929; *New York Graphic Magazine*, January 17, 1931; *Philadephia Record*, December 1, 1932; *Brooklyn Eagle*, April 17, 1931; *New York Times*, May 29, 1932. Clipping File, Buckminster Fuller Foundation, Los Angeles.

9. *Chicago Evening Post*, July 2, 1929.

10. "When We Live in Circles and Eat in Merry-go-rounds," *New Orleans Tribune*, June 21, 1932.

11. Fuller, *4-D*, p. 11.

12. Margaretha Jean Darnell, "Innovations in American Prefabricated Housing, 1860–1890," *Journal of the Society of Architectural Historians* 31 (1972), pp. 51–55; Charles Peterson, "Early American Prefabrication," *Gazette des Beaux-Arts*, series 6, vol. 33 (1948), pp. 37–46; James L. Garvin, "Mail-Order House Plans and American Victorian Architecture," *Winterthur Portfolio* 16 (1981), pp. 309–334.

13. "Housing VI: Solutions," *Fortune* 6 (July 1932), p. 61. The series was published in book form as *Housing America* (New York: Harcourt, Brace, 1932). The attribution to MacLeish (who was one of *Fortune*'s editors at the time) is in Arthur Mizener's *A Catalogue of the First Editions of Archibald MacLeish* (Folcroft, Pa.: Folcroft Library Editions, 1974; reprint of 1938 edition published by Yale University Library).

14. Edward A. Filene, *The Way Out* (Garden City, N.Y.: Doubleday, Page & Co., 1925), p. 180. See also Douglas Haskell, "Houses Like Fords," *Harper's* 168 (February 1934), pp. 280–298; D. Haskell, "The House of the Future," *New Republic* 66 (1931), pp. 344–345; Lewis Mumford, "The Flaw in the Mechanical House," ibid. 67 (June 1931), pp. 65–66.

15. Charles-Edouard Jeanneret (Le Corbusier), *Towards a New Architecture* (London: Architectural Press, 1948), pp. 210–247; originally published as *Vers une architecture* (Paris, 1923); first English translation by Frederick Etchells (London: Rodker, 1927).

16. Quoted in Thomas Hines's *Richard Neutra and the Search for a Modern Architecture* (New York: Oxford University Press, 1982), p. 78.

17. Hines, *Neutra*, p. 41.

18. John Coolidge, "The Modern House," *Atlantic Monthly* 159 (March 1937), pp. 286–296. See also Lewis Mumford, "Mass Production and the Modern House," *Architectural Record* 67 (January 1930), pp. 13–30.

19. Frank Chouteau Brown, "Chicago and Tomorrow's House?," *Pencil Points* 14 (June 1933), pp. 245–251; Dorothy Raley, *A Century of Progress Homes and Furnishings* (Chicago: Century Books, 1934).

20. William Keck, who assisted his older brother George during the design and construction of the fair houses, denies that the Dymaxion House had any influence on the configuration of the House of Tomorrow, though both Kecks were well acquainted with Fuller's work. A more important inspiration came from a nineteenth-century octagon house in the Kecks' home town of Watertown, Wisconsin. Author's interview with William Keck, Chicago, August 1982.

21. "The Future American Country House," *Architectural Record* 64 (1928), p. 417. Lescaze returned to the "home of tomorrow" theme in 1937 in two drawings for a streamlined "House of 2089," both now in the collection of the Avery Library at Columbia University. See David Gebhard and Deborah Nevins, *Two Hundred Years of American Architectural Drawing* (New York: Whitney Library of Design, 1977), p. 211. Gebhard and Nevins erroneously date these later efforts to 1927.

22. Quoted in Thomas M. Slade's "The 'Crystal House' of 1934," *Journal of the Society of Architectural Historians* 29 (1970), p. 350.

23. "The Prefabricated House Marches On," *Technology Review* 36 (1933–34), pp. 226–227.

24. "A Symposium on Prefabrication," *House and Garden* 68 (December 1935), pp. 70–71.

25. Robert S. Lynd and Helen Merrell Lynd, *Middletown: A Study in Modern American Culture* (New York: Harcourt, Brace, World, 1956), p. 107.

26. L. Rohe Walter, "Look Homeward, America!," *Review of Reviews and World's Work* 90 (October 1934), p. 27. For a similar scenario, see "Houses of the Future," *Scientific American* 147 (October 1932), p. 229.

27. "A Product of General Houses," *Architectural Forum* 57 (July 1932), p. 65.

28. "Mass-Produced Houses in Review," *Fortune* 7 (April 1933), pp. 52–57; "Housing VI: Solutions," ibid. 6 (July 1932), pp. 69, 104–108.

29. "Sears, Roebuck Boards the Prefabrication Bandwagon," *Architectural Forum* 63 (1935), p. 452.

30. "Sectional Motohome Makes Debut in Cellophane," *News-Week* 5 (April 13, 1935), p. 28.

31. Katharine M. Bissell, "The New American Home: An Interview with Robert W. McLaughlin, Jr., Architect," *Women's Home Companion* 62 (March 1935), p. 60.

32. "The House that Runs Itself," *Popular Mechanics* 63 (June 1935), pp. 805–806.

33. "Mass-Produced Houses in Review," p. 56.

34. "Machine for Living," *Business Week*, December 15, 1934, p. 9; "Houses, Inc.'s 'Motohome,'" *Architectural Forum* 62 (May 1935), pp. 508–510. Though Houses, Inc., failed, Gunnison went on to write a successful chapter in prefabrication history by forming Gunnison's Magic Homes, Inc., putatively the first company to produce houses on a moving assembly line. See Alfred Bruce and Harold Sandbank, *A History of Prefabrication* (New York: Arno, 1972; reprint of 1944 edition, issued as *Housing Research*, vol. 3), p. 64.

35. Hines, *Neutra*, p. 128.

36. "Steel Shell Houses," *Business Week*, March 30, 1935, p. 16.

37. John Sergeant, *Frank Lloyd Wright's Usonian Houses* (New York: Whitney Library of Design, 1976).

38. A. C. Shire, "Prefabricated Construction," *Housing Officials' Yearbook 1936* (Chicago: National Association of Housing Officials, 1936), pp. 140–142.

39. Miles Colean, *American Housing: Problems and Prospects* (New York: Twentieth Century Fund, 1944), p. 147.

40. Mumford, "The Flaw in the Mechanical Houses," p. 66.

41. Joseph L. Arnold, *The New Deal in the Suburbs: A History of the Greenbelt Town Program, 1933–1954* (Columbus: Ohio University Press, 1971); Paul L. Conkin, *Tomorrow a New World: The New Deal Community Programs* (Ithaca, N.Y.: Cornell University Press, 1959); Richard Pommer, "The Architecture of Urban Housing in the United States during the Early 1930s," *Journal of the Society of Architectural Historians* 37 (1978), pp. 235–264.

42. Robert W. Marks and R. Buckminster Fuller, *The Dymaxion World of Buckminster Fuller* (New York: Anchor, 1973), p. 23.

43. "Machine Made Steel Houses Envisioned as the Mass Product of the Future," *Iron Age*, June 2, 1932, p. 1217; R. T. Mason, "Steel Residences—Is This Market Appreciated?" *Steel* 87 (October 16, 23, 30, 1932), pp. 50–53, 55–59, 54–55.

44. See "Mass Produced Houses on Review," pp. 56, 83–84.

45. J. James and E. Wheeler, *Treasure Island: The Magic City* (San Francisco: Exposition Books, 1941), p. 150. A lineal descendant of the "House of Magic" was the "Carousel of Progress" sponsored by G.E. at Disneyland in the 1950s and the 1960s.

46. Quoted by T. S. Rogers, "A House that Forecasts the Home of Tomorrow," *American Architect*, March 1934, p. 3.

47. "Westinghouse House," *Business Week*, 3 March 1934, pp. 14–16; W. B. Courtney, "There's a Great Day Coming," *Collier's* 93 (January–June 1934), p. 60.

48. "The House that Works," *Fortune* 12 (October 1935), p. 59.

49. Siegfried Giedion, *Mechanization Takes Command* (New York: Norton, 1969; reprint of 1948 edition), pp. 532, 617–618; J. Normile and W. Adams, "Preview of Kitchen to Come," *Better Homes and Gardens* 21 (July 1943), pp. 38–40.

50. *House Beautiful* 85 (August 1943), pp. 48–49.

51. *Architectural Record* 94 (July 1943), p. 41.

52. Eric Hodgins, *Mr. Blandings Builds His Dream House* (New York: Simon and Schuster, 1946).

8

Skyscraper Utopias: Visionary Urbanism in the 1920s

Carol Willis

In the 1920s, Americans were more conscious than ever before that the city was the arena of their future. As the statistics of the 1920 census evidenced, the nation was for the first time predominantly urban, and all signs augured further centralization.[1] One response to this new urban identity was a heightened interest in the city of the future, particularly the skyscraper city. In newspapers, Sunday supplements, magazines, books, and movies, as well as in exhibitions in galleries, department stores, and expositions, Americans beheld fantastic images of future skyscraper utopias.

In October 1925, for example, thousands viewed an exhibition at the John Wanamaker department store entitled "The Titan City, a pictorial pageant of New York, 1926–2026."[2] Contrived both as publicity and as popular entertainment, the show featured murals of a spectacular skyscraper metropolis with colossal setback towers spaced at regular intervals and connected by multilevel transit systems, arcaded sidewalks, and pedestrian bridges at the upper floors. The harmonious avenues materialized in miniature in the store's main corridors, where model skyscrapers of fantastic shapes and colors encased the piers, creating a "Grand Canyon of the future."

The Titan City typified a new conception of the urban future that evolved in the 1920s—a modern metropolis of high density, advanced technology, and centralized planning. In contrast, most prognosticators around the turn of the century had foreseen a metropolis of giant, crowding towers—chaotic, congested, and teeming with technological gadgetry. In the early 1920s, however, these apprehensive projections of uncontrollable growth were supplanted by sanguine prophecies of an absolute urban order. During these years, a few farsighted architects envisioned an ideal city not simply transformed by technology but also rationalized by planning. Their optimistic visions of a modern skyscraper metropolis were quickly embraced by the general public and by many in the architectural profession. By the second half of the decade, this rationalized city of towers had become the new popular image of the urban future.

Although American utopianism had traditionally eschewed the city, these 1920s prophets advanced a nearly unanimous expression of urban optimism.[3] They shared a resolute faith that science and technology could solve all social problems, and they embraced the machine as man's liberator. But their faith in the inexorability of technological progress was also buttressed by a new confidence in their capacity to shape the urban future. Armed with the new planning principle of zoning and inspired by the recent revolutionary advances in engineering, transportation, and construction, they expected to build skyscraper cities that would be ordered, efficient, and hygienic—in short, everything that the contemporary city was not. Summoning this urban ideal in his 1930 book *The New World Architecture*, the critic Sheldon Cheney exhorted: "Let the vision be of a city beautiful, clean-walled, glowing with color, majestically sculptural, with a lift toward the skies; and let it be simple, convenient, sweet-running, airy, and light. . . . Thinking about it, visioning it, will make it come true some day."[4]

That the skyscraper should become the instrument of a new urban order was fitting, for by the 1920s the tall building had come to be perceived as a source of national pride and identity.[5] Whether viewed with awe or with apprehension, by critics or by admirers, skyscrapers were unanimously declared the quintessential expression of America in architecture. They were, as the architect Claude Bragdon proclaimed, "the symbol of the American spirit—restless, centrifugal, perilously poised."[6] Even Lewis Mumford, an implacable foe of tower cities, conceded: "There is, it is true, one universal and accepted symbol of our period in America: the skyscraper."[7] Both modern and distinctly American, skyscrapers seemed to confirm the nation's cultural maturity and, some asserted, her world supremacy in architecture.[8]

In these visionary schemes, high-rise buildings would house all aspects of urban life, including business, residence, recreation, and religion. Most would take the form of giant ziggurats with multiple setbacks, or of sheer towers, widely spaced for light and air. Traffic circulation would be facilitated by high-speed transportation systems, often linking the tops of the towers. Exploiting new materials and technologies, superblock structures would shoot up 100 stories and higher and would illumine the night sky with their electric beacons. The modernity of these towers envisioned in the 1920s, however, lay not so much in their colossal scale (many were in fact smaller than

Figure 1
"New York City in 1999," drawing by Biedermann. *New York World*, supplement, December 30, 1900.

those proposed in previous decades) as in their clean lines and simple, sculptural massing, as well as in the rejection of historicist ornament. More important, though, than the new forms or style of the towers was the idea of their disposition in a rational city plan.

To fully appreciate the radical change represented by the skyscraper-city visions of the 1920s, we must look back at popular images of the city of the future in the "pre-modern" period. Usually illustrated from bird's-eye perspectives, the drawings of this earlier period predicted a solid city, with every inch man-made. In these images one feels the compressed energy of the laissez-faire city—unleashed by new technologies such as steel-cage construction—exploding upward in titanic towers and imploding in denser concentrations of buildings, population, and enterprise. Besides the exaggerated heights and densities, there was also an emphasis on advanced transportation systems—multideck highways, aerial bridges, and rooftop platforms for airships.

A supplement to the *New York World* of December 30, 1900,[9] included an illustration by an artist named Biedermann depicting a 1999 dirigible view of Manhattan and the Brooklyn and New Jersey shores entirely built over with huge boxlike buildings, many with flat roofs for airship landings (figure 1). Lower Manhattan (which today is dominated by the twin towers of the World Trade Center) is, somewhat prophetically, shown with two pairs of giant towers. A host of new bridges link the city with the opposite shores, though the river remains a major traffic route. Another fanciful Biedermann drawing, this one from 1916, illustrates the upper level landscape of an imaginary metropolis in which the roofs of skyscrapers provide terraces for a variety of fantastic transport systems. Multiple levels of serpentine highways and rails wind between buildings like a Coney Island roller coaster. Above, great metal truss structures support platforms for bizarre flying machines.[10]

Perhaps the most popular and perennial prophecy of the New York of the future is one that appeared in several updated versions (from 1908 to 1915) in a well-known picture book of city landmarks, *King's Views of New York.*[11] Imagining a vista up Broadway in about 1930, this perspective shows a sheer-walled canyon of great office blocks, which dwarf the Singer Building—once the world's tallest. Bridges spring from one rooftop to the next or tunnel through upper floors, and the sky swarms with dirigibles and airplanes. A caption seconds the impression of chaotic concentration, describing this vision of the "Cosmopolis of the Future" as "a weird thought of the frenzied heart of the world in later times, incessantly crowding the possibilities of aerial and inter-terrestrial construction."[12]

This popular image of a vertiginous city of towers, overtaken by rampant technology, was satirized by many early-twentieth-century illustrators. In one of his drawings for *Life* in 1912, for example, the cartoonist Harry Grant Dart showed a future baseball game played in a skyscraper stadium, with spectators perched precariously on the roofs of surrounding towers. Winsor McCay, the creator of "Little Nemo" and other early comic strips, often drew giant agglomerations of teetering towers. Many other examples could be cited, for in the early twentieth century illustrators and cartoonists dominated the field of future-gazing. Little has been written about these artists or their sources, although their method of projection can be compared to that of their contemporary H. G. Wells (who, as the most popular and prolific writer of futuristic fiction of the period, was himself a pervasive source of ideas for the city of tomorrow). Wells described the method he had employed in his prophecy of London in the twenty-first century, *When the Sleeper Awakes* (1899), as "enlarging the present." Beginning with the present-day city, he applied the mathematical "rule of three"—he multiplied everything threefold. "In that fashion," he said, "one got out a sort of gigantesque caricature of the existing world, everything swollen up to vast proportions and massive beyond measure."[13]

Figure 2
"King's Dream of New York," drawing by Harry M. Pettit. *King's Views of New York*, 1908 edition.

Indeed, "enlarging the present" aptly characterized the method by which most of these pre-1920s prophets arrived at their conceptions. Their chimerical images were simply extrapolations of the contemporary city and its problems, not proposals for alternatives. These artists assumed that technological progress and the competitive forces of the laissez-faire economy would inevitably produce such scenes. Their drawings therefore betray a certain ambivalence; although they were often humorous and high-spirited, they also carried the sobering intimation that this supercharged city could not be controlled.

In the early years of the century, then, the popular image of the future city was principally the creation of artists and illustrators. (By "popular image" I mean the image that was most frequently advanced and most widely accepted by the general public.) It is a phenomenon of American futurism that, at different times, different professions have been the principal inventors of popular imagery. From the late 1880s until about 1900, inspired in particular by the visions of Edward Bellamy, utopian novelists predominated. From the turn of the century until about 1920, as we have seen, professional illustrators and cartoonists created the most engaging predictions of the future city, while most architects and planners adhered to an urban ideal modeled on great European cities of the past. In the 1920s, however, architects became the chief oracles of the urban future and, through the influence of a few prescient designers, invented a new vision of the skyscraper city.

The 1920s were pivotal years in the changing conception of the city, present and future. Summarizing the shifts in American society from the prewar to the postwar period, the historian William Leuchtenberg has noted: "In 1916, Americans still thought to a great extent in terms of nineteenth-century values of decentralization, competition, equality and agrarian supremacy of the small town. By 1920 the triumph of the twentieth century—centralized, industrialized, secularized and urbanized—while by no means complete, could clearly be foreseen."[14] In architecture, this increasing urbanization was answered by a growing professionalism in city planning and in the development of related disciplines of urban studies, particularly urban sociology. In addition, the new legal concept of zoning offered architects an unprecedented tool with which to control urban growth.

The importance of zoning as a catalyst for a new conception of the city of the future cannot be underestimated. Indeed, it can be argued that the concept of the city rationalized by planning was a direct outgrowth of the country's first comprehensive zoning law, the New

York City statute of 1916.[15] Enacted after nearly two decades of debate, this legislation was designed by its original proponents principally to protect public health and safety and property values. In the 1920s, however, a "second generation" of zoning advocates emerged. These were skyscraper architects who, analyzing the limits of the code, found a new formal inspiration, which they applied first to designs for individual buildings and then to their conceptions of the city. Of particular importance in this development was the idea of the "zoning envelope"—a restriction on the maximum legal volume of a tall building, requiring that it be stepped back in ziggurat fashion at specified levels. In this formula a few perspicacious architects discerned the basis for a new stylistic expression for the skyscraper, an aesthetic of simple, sculptural masses and subordinated ornament which they declared both modern and distinctly American. An enlargement of this setback formula also created a new convention in many visionary prophecies: the "superblock," a giant stepped-back tower rising over a multiblock base. The setback formula and the superblock suggested a future urban topography of rationally spaced towers.

Descriptions of the impact of zoning on the future city in the writings of architects and critics of the period often resembled paeans to progress. In his 1928 book *American Architecture of Today*, Harvard architecture professor G. H. Edgell hailed zoning as "the first great instrument of control, working for the good of science, humanity, and of art."[16] The architect A. N. Rebori found the Chicago zoning law, passed in 1922, "a strong stimulus to the creative mind," and predicted that it would produce "a city of towers, shimmering skyward, a symbol of man's enlightenment, when science and art locked hands in lasting tribute to the age in which we live."[17]

Earlier in the century, during the years of vigorous laissez-faire growth and the philosophy of "rugged individualism," comprehensive control over an entire city—particularly the central business district, with its high land values—had been virtually inconceivable to most architects and planners.[18] But with the new tool of zoning, architects began to assert a new sense of efficacy in their capacity to plan the modern metropolis, and their drawings communicated this new confidence. Their "postzoning mentality," as it may be termed, contrasted strikingly with the "prezoning mentality" of the earlier prophets of the future city, who had assumed that urban growth could not be regulated. The antithetical nature of these two concepts of the city is immediately apparent when one contrasts the images of the chaotic and crowding towers of the prezoning period, such as the *King's Views* cover shown in figure 2 above, with the serene and orderly urban vistas of Francisco Mujica (figure 4) and Hugh Ferriss (figure 6).

By far the most significant roles in awakening the profession to the formal and planning implications of the new zoning laws were played by the skyscraper architect Harvey Wiley Corbett and the architectural delineator Hugh Ferriss. Early in 1922, they began a theoretical study of the requirements of the New York setback ordinance and of how these restrictions could most profitably be applied to a viable commercial structure. Their collaboration spawned a series of articles in which, with escalating optimism, they proclaimed the propitious effects of zoning on skyscraper design and on the future metropolis.[19] Ferriss also created the famous series of renderings of the "four stages of the building envelope," in which he gave the legal formula for the setback an almost iconic identity. Widely published and exhibited throughout the 1920s, these drawings were of unparalleled importance in impressing other architects with the power, beauty, and inchoate modernity of the unornamented setback form.

Ferriss made a striking visual and verbal equation of the simple setback mass and the modern style in an article entitled "The New Architecture" in the *New York Times* of March 19, 1922. "We are not contemplating the new architecture of a city," he declared, "we are contemplating the new architecture of a civilization."[20] In a second piece six months later, he predicted that within a generation American cities would be transformed by the new zoning laws, and he accompanied his forecasts with fantastic drawings of future urban vistas.[21]

Inspired by Ferriss's grandiloquent visions, Corbett soon began to echo his colleague's optimism. In an interview he gave for a *New York Times* article that appeared on April 29, 1923, he called the zoning-inspired setback style "utterly new and distinctive" and predicted that it would affect the architecture of the whole world.[22] In another piece, Corbett proclaimed zoning "a new idea in city building" and predicted: "The new type of city with its innumerable spires and towers and domes setback from the cornice line, will provide a fascinating vision—all the novelty in the world brought under a larger scheme."[23]

Corbett's ideas were expanded in another collaboration with Ferriss, the "Titan City" exhibition described above. The fantastic visualizations of the Titan City were really a summary of many of the professional speculations and pet projects that had occupied the two architects for several years. The zoning-envelope studies were featured in a series of 12-foot monochromatic paintings by Ferriss and further developed in designs for colossal towers based on the setback principle. Ferriss also enlarged several penthouse panoramas of the future city which

Figure 3
Harvey Wiley Corbett, mural from Titan City exhibition, 1925. Robert W. Chanler, delineator.

he had previously published in earlier articles and had exhibited at his first one-man show at the Anderson Galleries in April 1925. Corbett's earlier work for the Russell Sage Foundation on traffic separation provided the concept behind another series of views of grand avenues and terraced promenades, rendered by the artist Robert Chanler and his staff (figure 3). The show also portrayed ideas by other designers which had recently been published in the popular press. Ferriss's mural of apartments on bridges, for example, was a variation on Raymond Hood's "bridge homes," which had appeared in the *New York Times* in February 1925.[24] Airplane landing platforms built on the roofs of skyscrapers were copied from schemes developed by some of Corbett's students for a Beaux Arts Institute of Design competition of 1925.

The Titan City exhibition helped spark a brilliant efflorescence of visionary urbanism. Indeed, in the later 1920s both professional and public interest in cities of the future reached unprecedented heights. Many different predictions and proposals were advanced during these years of booming prosperity. For convenience, they may be divided into four types: The most frequent approach envisaged a city plan of regularly spaced setback towers, usually connected by aerial highways. In the second type, skyscrapers were clustered in nucleated centers. A third variation proferred isolated towers surrounded by open space. The megastructure—an entire city contained within a single building—constituted a minor fourth type.

Figure 4
Francisco Mujica, "The City of the Future: Hundred-Story City in the Neo-American Style." Mujica, *History of the Skyscraper* (New York: Da Capo, 1977, reprint of 1929 edition), plate CXXIV.

The vision of orderly avenues of pyramidal towers was initially promulgated by Corbett and Ferriss, but it subsequently appeared in a number of projects by other designers. One particularly regimented version was the "hundred-story city in the Neo-American style" (figure 4) proposed by Francisco Mujica in the conclusion of his 1929 book *History of the Skyscraper.*[25] Mujica, a Mexican archeologist and architect, believed that the pre-Columbian stepped pyramid offered an appropriate formal precedent and a decorative vocabulary suited to the requirements of the New York zoning law and meaningful as an indigenous "American" style. He envisioned regular rows of nearly identical setback towers, connected by elevated pedestrian bridges and layered levels of traffic below. Even though Mujica's perspective drawing seems relentlessly rational, his declared motives were idealistic and humanistic. In his book's brief text he exhorted: "Let us meet the future halfway, be modern without breaking with the past—let us be practical, yet set aside a few moments each day for dreaming . . . and thus work for the development of the modern city, giving it fully the mechanical and practical stamp of our century, yet not forgetting in our planning that each part of this gigantic human machine has a heart capable of soaring high and loving. . . ."[26]

Figure 5
Set for the 1930 Fox film *Just Imagine*. Stephen Gooson and Ralph Hammeras, designers.

A Hollywood version of a concept similar to Mujica's hundred-story city was created as a set for the 1930 Fox film *Just Imagine*.[27] Billed as "the first science-fiction musical," the movie treated audiences to a fantasy visit to New York in 1980. In an old Army blimp hangar, set designers Stephen Gooson and Ralph Hammeras constructed a 225-by-75-foot model of a glittering metropolis (figure 5). Lofty towers of up to 250 stories, nine levels of multilane traffic systems complete with moving cars, personal airplanes, and aerial traffic cops were all features of this minutely detailed set. Built to a scale of 1/4 inch to 1 foot, the tallest model (40 feet) represented a building almost 2,000 feet high. In many aspects, this set foreshadowed the extravagant models and dioramas created by industrial designers for the 1939 New York World's Fair, particularly the Democracity of Henry Dreyfuss and the Futurama of Norman Bel Geddes.

Many proposals by serious designers equaled the fantasies of *Just Imagine*. In 1929 the transportation engineer Robert Lafferty proposed an elaborate multilevel traffic system that would include "skyway" travel. This was a scheme for a continuous bridge-highway to be suspended between the rising shafts of setback towers, called "station pylons." Lafferty argued the virtues of his system, claiming that "the 6″ to 8″ cables, and light and strong structure of Airways 200′ in the air will cast but little shadow," and that "as this system will minimize noise and vibration, the beauty and attractiveness is obvious."[28] Another traffic engineer, John A. Harriss, proffered a plan for the "American Multiple Highway." In an article published in the planners' journal *The American City* in June 1927, he pictured a scheme for a network of multilane highways stacked six deep across Manhattan.[29]

In his 1929 book *The Metropolis of Tomorrow*, Hugh Ferriss created the 1920s' most complete and compelling vision of the future city.[30] His scheme was a hybrid of widely spaced setback towers and centers of concentrated development (figure 6). Like a Renaissance ideal city, Ferriss's metropolis possessed a strong geometric plan—two triangles forming a six-pointed star, inscribed within a circle—and a hierarchical formal arrangement of structures that was also symbolic. The city was divided into three major zones—Business, Science, and Art—each dominated by a great setback "tower-building" rising 1,000 feet or higher on a base of four to eight full city blocks. Spaced at half-mile intervals, these "primary centers" were to be built over the intersections of 200-foot-wide avenues and to serve as "express stops" for the efficient highway system. Other secondary centers were positioned in relation to this principal trinity—for example, at the midpoint between the Art and Business sectors rose the center for the Applied and Industrial Arts, and between Art and Science stood the tower of Philosophy. Although this absolute hierarchy may today seem oppressive or authoritarian, to Ferriss it signified harmony and humanism. He asserted in the conclusion of his book that this city would be "populated by human beings who value mind and emotion equally with the senses, and have therefore disposed their art, science, and business centers in such a way that all three would participate equally in the government of the city."[31]

A less idealized scheme for nucleated centers of development was a 1929 project by Raymond Hood called "Manhattan 1950."[32] As a winner of the competition for the *Chicago Tribune* tower, the designer of the *Daily News* building in New York, and a partner in the Rockefeller Center design team, Hood was one of the country's most celebrated skyscraper architects. In "Manhattan 1950," he prophesied that skyscrapers would develop in clusters across the island, and he presented a plan for more than twenty tentacular bridges on which would be built luxury apartment towers, each of which might house 10,000 to 50,000 residents (figure 7).

Figure 6
Hugh Ferriss, "Looking West from the Business Center." Ferriss, *The Metropolis of Tomorrow* (New York: Ives Washburn, 1929), p. 121.

Figure 7
Raymond Hood, "Manhattan, 1950," aerial view of skyscraper bridges. Proposal, Annual Exhibition, Architectural League of New York, 1930.

In a series of projects proposed a few years earlier, Hood had championed another type of city plan: isolated towers surrounded by open space. The earliest of these schemes, which appeared in the *New York Times Magazine* in December 1924, was illustrated with a Ferriss drawing entitled "City of Needles."[33] Hood proposed pencil-thin towers that would rise over a narrow base to heights of 1,000 to 1,400 feet. He maintained that concentrating activities in free-standing towers would free the surrounding space for parks or traffic circulation, and he elaborated this concept of a city of towers, as he called it, in several later projects that were widely publicized in architectural circles.[34]

An entire city within a single giant skyscraper—what today we call a megastructure—was a minor category of futuristic speculation during the 1920s. Illustrations of megastructural buildings often appeared in the pages of science-fiction magazines, where the artist Frank R. Paul

Figure 8
Hugh Ferriss's rendering for a proposal for towers spaced 500 feet apart. First published in Hood's "New York Skyline Will Climb Much Higher," *Liberty*, April 10, 1926, p. 19.

invented scores of superscale cities.[35] Ironically, though, the exceedingly dense and oddly eclectic cities of Paul and other illustrators today seem retardataire in comparison with those of Ferriss or Mujica. Paul's futuristic fantasies perpetuated the earlier, prezoning mentality. Though his prophecies were obsessed with advanced technology and colossal scale, they were not rationalized by planning. A project by Lloyd Wright (son of Frank) published in the *Los Angeles Examiner* in November 1926 rivaled the work of science-fiction artists.[36] Wright proposed a 1,000-foot tower that would cover an area of 40 acres at the center of a 20-mile square of farms, forests, and parks. This single structure (figure 10) would contain all the industrial, commercial, and residential functions of a community of 150,000. Although Wright's schemes owed much to his father's formal and planning ideas, this drawing was purposely given a comic-book character that undermined its seriousness.

All these various visionary schemes proffered a new vision of the clean, efficient, rational skyscraper city. This ideal of the modern metropolis wedded the fascination with technology of the earlier, prezoning predictions with a new optimism and confidence about man's ability to control and direct future urban growth. The conditions that encouraged this optimism were many and complex, but three fundamental factors should be noted: the seminal influence of zoning, already discussed; an enthusiasm for new materials and technologies; and a fervent faith in progress.

The impact of technology on these visionary schemes seems almost too obvious to elaborate. Their images are intoxicated with technology—the attentuated towers, the obsession with high-speed transportation, the aerial perspectives, the dramatic night views radiating man-made light. None of these enthusiasms were new, however. Dreams of 1,000-foot towers had appeared frequently in architectural and engineering journals in the early twentieth century, and such towers were considered quite buildable. What was novel about the super skyscrapers proposed in the 1920s was not their height, but their simplicity of form, the deemphasis of ornament, and their placement in a rationalized city plan. With the important exception of glass for facades, most of the building materials projected for the future towers were already available during the 1920s. Schemes for multilevel mass transit systems had been an obsessive interest since the middle of the nineteenth century, as had been air travel (first by balloon or dirigible, then, after the Wright brothers, by heavier-than-air machines). The airplane and the automobile became part of the essential iconography of the future city not simply for their wondrous technology

Figure 9
Frank R. Paul, illustrations for Earl L. Bell's "The Moon Doom," *Amazing Stories*, Winter 1928.

Figure 10
Lloyd Wright, proposal for 1,000-foot tower. *Los Angeles Examiner*, November 26, 1926.

(as had been the case in the early twentieth century) but for their promise of personal freedom. With the automobile, this freedom was already being integrated into a new American lifestyle; the airplane seemed to offer even greater liberation. Highways and airports, often on or spanning the roofs of skyscrapers, became prominent design features of the cities of the future.

Indivisibly bound to the enthusiasm for technology was the traditional American faith in the idea of progress. "There was no doubt about the future in America," stressed historian and social critic Charles Beard, writing in 1928; "the most common note of assurance was the belief in unlimited progress."[37] To most twentieth-century Americans the word *progress* signified technological and material advancement, but social progress was a corollary. Describing the idea of progress in contemporary American society, Beard asserted that "mankind, by making use of science and invention can progressively emancipate itself from plagues, famines, social disasters, and subjugate the material forces of the good life—here and now."[38] Emphasizing the collective good, Beard closely associated progress with democracy and social amelioration.

The urban visions of the 1920s incarnated the dreams of a democratic, technological utopia. The writings of Ferriss, Mujica, Cheney, and others clearly announced their aspiration to create a city of harmony, beauty, and efficiency that would benefit all its citizens. They were confident that such a radical reshaping of the city could be realized through progress and planning if people were willing to place restrictions on the rights of private property and to institute centralized controls. That their schemes seem naive, misguided, or worse from today's critical perspective is not the concern here; rather, the aim is to recognize how, in the context of their times, these dreams seemed realizable.

The visionary architects' belief in centralized planning concorded with a growing consensus in many areas of American society that planning, and not "rugged individualism," would necessarily be the way of the country's future. Some historians have linked this conviction to the successful mobilization of industry during World War I, the nation's first major attempt at government control over the economy. In his revisionist appraisal of the 1920s, historian Ellis Hawley has argued that the effectiveness of the wartime mobilization offered a model for peacetime management that became a central theme of postwar policy. Hawley breaks with the conventional view of the 1920s as a period of both competition and conservatism in business, and instead emphasizes the emergence of a new ideal of government by "managerial and bureaucratic institutions, led by organizational and technical experts, and dedicated to solving the nation's problems and advancing the common good."[39] Though they rarely even acknowledged governmental or social systems—the traditional concern of utopias—the proposals for skyscraper cities should be understood in the context of the period's ideal of managerial capitalism.

The American visionary architects of the 1920s believed that change would be evolutionary, not revolutionary. True sons of their decade of bullish prosperity, they saw no conflict between utopianism and capitalism. If action were to be taken to set rational and reasonable guidelines for future urban growth, they assumed, the irrepressible vitality of the capitalist system would eventually materialize their ambitious designs.

The most prominent and active theorists of the modern skyscraper city were the architects Corbett, Ferriss, and Hood. The traditional definition of *theory* must be qualified, however, for in the modern movement in architecture theory and ideology became virtually inseparable. These American visionaries were not radicals or ideologues; their writings contained no explicit polemics, like those of the Bauhaus or the Russian Constructivists. Their philosophy can be thought of as a sort of "passive modernism," as opposed to the "active modernism" of such avant-garde Europeans as Le Corbusier, who offered the ultimatum "architecture or revolution."

The sanguine prophecies of the American visionaries were not without critics or alternative philosophies. Most contemporary criticism was concentrated on two points: that the dense, intense skyscraper cities were inhuman in scale, and that they were economically impracticable. Some viewed them as malignant symbols of capitalist greed. Lewis Mumford, in his *New Republic* review of the Titan City exhibition, railed that "innumerable human lives will doubtless be sacrificed to Traffic, Commerce, Properly Regulated and Zoned Heights on a scale that will make Moloch seem an agent of charity."[40] Mumford was a leading spokesman of the regionalist movement, a planning philosophy that won many influential adherents in the 1920s and the 1930s and became, in effect, the intellectual establishment of the American planning profession. In addition to their basic tenet that comprehensive planning must be coordinated by region rather than fragmented in separate municipalities, the regionalists held a deep ideological commitment to decentralizing cities. Their chief interest lay not with the macro-picture of collective public good, but rather with the individual home and the neighborhood unit as the ideal of community.

The regionalists, in the end, proved to be the most accurate prognosticators of America's real (suburban) future, and as a result they have received more serious attention from historians than the prophets of skyscraper utopias. Unfortunately, the old polemics of the centralist-decentralist debate of the 1920s have carried over into modern histories of American planning (most of which have been written by partisans of the regionalist philosophy) and have colored the authors' interpretations of the visionary schemes.[41] Charges that these architects were naive with regard to economic and social issues, and accusations of megalomania, may be well founded. In addition to these points, however, many recent critics, writing from the perspective of the current widespread disillusionment with the modern movement, have characterized the centralized supercities of the 1920s as authoritarian, oppressive, monotonous, and sterile. Yet, as we have seen, their designers intended exactly the opposite; to them, the vision of urban order was democratic, liberating, harmonious, and hygienic.

In the 1920s, Americans were adjusting to their country's new identity as an urban nation and a preeminent world power. During this period of accelerated growth and social change, the prophecies of rationalized skyscraper cities fixed a confident image of America's urban future. Skyscrapers were perceived as the symbol of America's progress and of her ascendancy on the world stage. "At the climax of a nation's development, it is the impulse of a colossal work which has set its stamp upon the acknowledged type of the time and has caused that work to be characteristic of the nation," wrote one critic in a 1925 article entitled "America's Titanic Strength in Architecture."[42] The point was well taken, for the dreams of future skyscraper cities were a quintessentially American vision. They at once signaled the demise of the persistent national inferiority complex in the face of European culture and offered, in the New World, a prefiguration of the global future. In assuming that technology could be tamed, the city planned, and the future designed for the benefit of mankind, the visionary architects of the 1920s became the masters of the machine-age metropolis and the creators of America's first modern conception of the city as utopia.

Notes

1. The 1920 census is often cited by historians as a watershed in America's metamorphosis from a rural to an urban society, even though, as is frequently noted, *urban* was applied to communities as small as 2,500. For a brief discussion of the significance of the census figures see John Garraty, *The American Nation*, third edition (New York: Harper and Row, 1975), pp. 684–685.

2. For a description of this little-known exhibition, see my article "The Titan City," *Skyline*, October 1982, pp. 26–27.

3. An exception was a body of utopian literature published from the late 1880s to about 1900. Inspired principally by the success of Edward Bellamy's *Looking Backward*, a number of utopian novels of this period were set in cities. Although these fictional cities were sometimes described in detail, and very occasionally illustrated, the city plans and the styles of their buildings always remained a secondary consideration; social and governmental organization was primary. For discussions of late-nineteenth-century utopian literature see chapter 6 of this volume and Kenneth Roemer, *The Obsolete Necessity* (Kent, Ohio: Kent State University Press, 1976).

4. Sheldon Cheney, *The New World Architecture* (New York: Longmans, Green, 1930), pp. 398–399.

5. Before the 1920s, attitudes toward the skyscraper expressed in popular and professional journals were often negative or apprehensive. In the 1920s, though there were still many critics, the praise for tall buildings was stronger than ever before. Stanley Andersen has discussed these changing attitudes in his Ph.D. dissertation, American Ikon: Responses to the Skyscraper 1875–1934 (University of Minnesota, 1960).

6. Claude Bragdon, "The Shelton Hotel, New York," *Architectural Record* 58 (July 1925), p. 1.

7. Lewis Mumford, "American Architecture To-day," *Architecture* 58 (October 1928), p. 189.

8. For example, in the new concluding chapter of the 1927 edition of his popular book *The Story of Architecture in America* (New York: Norton), Thomas E. Tallmadge stressed two points: "the discovery of the long-sought solution of the problem of the sky-scraper" and "our recently achieved world supremacy in architecture" (p. 290).

9. See "New York City as It Will Be in 1999; Pictorial Forecast of the City," supplement to *New York World*, December 30, 1900.

10. I do not know the original source of this drawing. It is reprinted on p. 57 of David Kyle's *Pictorial History of Science Fiction* (London: Hamlyn, 1976). The drawing may well date from before 1916, the date given by Kyle.

11. The first of the *King's Views* images was drawn for the 1908 edition by Harry M. Petit, who also contributed many other renderings of real city buildings to the volume. Updated versions with airplanes, aerial highways, etc. were developed by the illustrator Richard Rummell for the editions of 1911–12 and 1915. All three versions are included in *King's Views of New York 1886–1915*, Moses King, ed. (New York: Arno, 1974). These *King's Views* plates were often reprinted in other publications; this extended their influence and popularity. For example, Erich Mendelsohn included one in his 1929 book *Russland, Europa, Amerika*.

12. *King's Views*, 1908 edition.

13. Wells described his method of forecasting in *The Future in America* (New York: Harper and Bros., 1906), pp. 11–12, quoted in Bernard Bergonzi, *The Early H. G. Wells* (Manchester University Press, 1961), p. 144.

14. William Leuchtenberg, *The Perils of Prosperity* (University of Chicago Press, 1958), p. 43.

15. I have discussed the influence of zoning on the new idea of the skyscraper city in "Zoning and Zeitgeist: The Skyscraper City in the 1920s," *Journal of the Society of*

Architectural Historians (forthcoming). For background on the New York zoning law, see Mel Scott, *American City Planning Since 1890* (Berkeley: University of California Press, 1969).

16. G. H. Edgell, *American Architecture of Today* (New York: Charles Scribner & Sons, 1928), p. 356.

17. A. N. Rebori, "Zoning Skyscrapers in Chicago," *Architectural Record* 58 (July 1925), p. 90.

18. A few architects in the late nineteenth century and the early twentieth century did offer proposals for centralizing planning and regulating the rise of the skyscraper; Louis Sullivan and Charles R. Lamb are two. Yet none of their proposals were seriously received by contemporaries, because government control over a whole city was considered impossible.

19. From 1922 to 1924, Ferriss and Corbett collaborated on several articles, some published jointly, others separately. In addition, other authors featured their ideas or drawings. The most forceful of these articles are cited in the text or in notes below. These articles constitute virtually the whole of the publications during these years on the subject of the influence of zoning.

20. Hugh Ferriss, "The New Architecture," *New York Times*, March 19, 1922, section 4, p. 8.

21. Hugh Ferriss, "Civic Architecture of the Immediate Future," *Arts and Decoration* 18 (November 1922), p. 13.

22. Interview with Harvey Wiley Corbett in "The Coming City of Setback Skyscrapers," *New York Times*, April 29, 1923, section 4, p. 5.

23. H. W. Corbett, "What the Architect Thinks of Zoning," *American Architect* 125 (February 13, 1924), p. 150.

24. Orrick Johns, "Bridge Homes—A New Vision of the City," *New York Times*, February 22, 1925, section 4, p. 5. (Illustrated by Hugh Ferriss.)

25. Francisco Mujica, *History of the Skyscaper* (New York: Da Capo, 1977, reprint of 1929 edition).

26. Ibid., pp. 52–53.

27. Some information on *Just Imagine* has been reprinted in Tim Onosko's *Wasn't the Future Wonderful?* (New York: Dutton, 1979), pp. 42–43. The film department of the Museum of Modern Art has a print of *Just Imagine*, which can be viewed by special appointment.

28. Robert C. Lafferty, *The Lafferty Plan: Traffic, Transit, Freight* (New York: Culture Press, 1929).

29. "And This? Dr. John A. Harriss Proposes Six-Deck Streets," *American City* 36 (June 1927), pp. 803–804.

30. Hugh Ferriss, *The Metropolis of Tomorrow* (New York: Ives Washburn, 1929).

31. Ibid., p. 140.

32. Hood's visionary scheme was exhibited at the annual exhibition of the Architectural League of New York in 1930 and thereafter published in several places, including *Creative Art* 9 (August 1931), pp. 160–161.

33. Orrick Johns, "Architects Dream of a Pinnacle City," *New York Times*, December 28, 1924, section 4, p. 10.

34. I discuss the dating of and the influences on Hood's numerous visionary proposals for skyscraper cities in "Towering Cities," *Skyline*, July 1982, pp. 10–11. This piece

is a review of a book by Robert A. M. Stern with Thomas P. Catalano, *Raymond Hood* (New York: Institute of Architecture and Urban Studies and Rizzoli, 1982).

35. Frank R. Paul was the favorite illustrator of Hugo Gernsback, the famous publisher of early science-fiction magazines. His drawings appeared regularly in *Amazing Stories*; most date from the mid 1920s through the 1930s.

36. Lloyd Wright's drawing, accompanied by a brief descriptive article, appeared in the *Los Angeles Examiner*, November 26, 1926, Calendar section, p. 2.

37. Charles A. Beard and Mary R. Beard, *The Rise of American Civilization*, revised edition (1928; New York: Macmillan, 1944), p. 831.

38. Beard, "The Idea of Progress," in *A Century of Progress*, Charles A. Beard, ed. (New York: Harper and Bros., 1932), pp. 6–7.

39. Vincent Carosso, introduction to *The Great War and the Search for Modern Order* by Ellis W. Hawley (New York: St. Martin's, 1979), p. xv.

40. Lewis Mumford, "The Sacred City," *New Republic* 45 (January 27, 1926), pp. 270–271.

41. The best treatments of the history of American planning of this period are Mel Scott's *American City Planning Since 1890* (Berkeley: University of California Press, 1969) and Francesco Dal Co's "From Parks to Region: Progressive Ideology and the Reform of the American City," in Giorgio Ciucci et al. (eds.), *The American City: From the Civil War to the New Deal* (Cambridge, Mass.: MIT Press, 1979). Both of these authors express strong sympathies for regionalist philosophy—which they discuss at length—but neither displays an understanding of the idealism that motivated many of the centralists' city-of-the-future schemes.

42. C. H. Walker, "America's Titanic Strength Expressed in Architecture," *Current History* 25 (January 1925), p. 555.

9

An Unforeseen Revolution: Computers and Expectations, 1935–1985

Paul Ceruzzi

The "computer revolution" is here. Computers seem to be everywhere: at work, at play, and in all sorts of places in between. There are perhaps half a million large computers in use in America today, 7 or 8 million personal computers, 5 million programmable calculators, and millions of dedicated microprocessors built into other machines of every description.

The changes these machines are bringing to society are profound, if not revolutionary. And, like many previous revolutions, the computer revolution is happening very quickly. The computer as defined today did not exist in 1950. Before World War II, the word *computer* meant a human being who worked at a desk with a calculating machine, or something built by a physics professor to solve a particular problem, used once or twice, and then retired to a basement storeroom. Modern computers—machines that do a wide variety of things, many having little to do with mathematics or physics—emerged after World War II from the work of a dozen or so individuals in England, Germany, and the United States. The "revolution," however one may define it, began only when their work became better known and appreciated.

The computer age dawned in the United States in the summer of 1944, when a Harvard physics instructor named Howard Aiken publicly unveiled a giant elecromechanical machine called the Mark I. At the same time, in Philadelphia, J. Presper Eckert, Jr., a young electrical engineer, and John Mauchly, a physicist, were building the ENIAC, which, when completed in 1945, was the world's first machine to do numerical computing with electronic rather than mechanical switches.

Computing also got underway in Europe during the war. In 1943 the British built an electronic machine that allowed them to decode intercepted German radio messages. They built several copies of this so-called Colossus, and by the late 1940s general-purpose computers

ENIAC, ca. 1947. The plugboards by which the computer was programmed can be seen at left.

were being built at a number of British institutions. In Germany, Konrad Zuse, an engineer, was building computers out of used telephone equipment. One of them, the Z4, survived the war and had a long and productive life at the Federal Technical Institute in Zurich.

These machines were the ancestors of today's computers. They were among the first machines to have the ability to carry out any sequence of arithmetic operations, keep track of what they had just done, and adjust their actions accordingly. But machines that only solve esoteric physics problems or replace a few human workers, as those computers did, do not a revolution make. The computer pioneers did not foresee their creations as doing much more than that. They had no glimmering of how thoroughly the computer would permeate modern life. The computer's inventors saw a market restricted to a few scientific, military, or large-scale business applications. For them, a computer was akin to a wind tunnel: a vital and necessary piece of apparatus, but one whose expense and size limited it to a few installations. For example, when Howard Aiken heard of the plans of Eckert and Mauchly to produce and market a more elegant version of the ENIAC, he was skeptical. He felt they would never sell more than a few of them, and he stated that four or five electronic digital computers would satisfy all the country's computing needs.[1] In Britain in 1951, the physicist Douglas Hartree remarked: "We have a computer here in Cambridge; there is one in Manchester and one at the [National Physical Laboratory]. I suppose there ought to be one in Scotland, but that's about all."[2] Similar statements appear again and again in the folklore of computing.[3] This perception clearly dominated early discussions about the future of the new technology.[4] At least two other American computer pioneers, Edmund Berkeley and John V. Atanasoff, also recall hearing estimates that fewer than ten computers would satisfy all of America's computing needs.[5]

By 1951 about half a dozen electronic computers were running, and in May of that year companies in the United States and England began producing them for commercial customers. Eckert and Mauchly's dream became the UNIVAC—a commercial electronic machine that for a while was a synonym for *computer*, as *Scotch Tape* is for cellophane tape or *Thermos* is for vacuum bottles. It was the star of CBS's television coverage of the 1952 presidential election, when it predicted, with only a few percent of the vote gathered, Eisenhower's landslide victory over Adlai Stevenson. With this election, Americans in large numbers suddenly became aware of this new and marvelous device. Projects got underway at universities and government agencies across the United States and Europe to build computers. Clearly, there was a demand for more than just a few of the large-scale machines.

But not many more. The UNIVAC was large and expensive, and its market was limited to places like the U.S. Census Bureau, military installations, and a few large industries. (Only the fledgling aerospace industry seemed to have an insatiable appetite for those costly machines in the early years.) Nonetheless, UNIVAC and its peers set the stage for computing's next giant leap, from one-of-a-kind special projects built at universities to mass-produced products designed for the world of commercial and business data processing, banking, sales, routine accounting, and inventory control.

Yet, despite the publicity accorded the UNIVAC, skepticism prevailed. The manufacturers were by no means sure of how many computers they could sell. Like the inventors before them, the manufacturers felt that only a few commercial computers would saturate the market. For example, an internal IBM study regarding the potential market for a computer called the Tape Processing Machine (a prototype of which had been completed by 1951) estimated that there was a market for no more than 25 machines of its size.[6] Two years later, IBM developed a smaller computer for business use, the Model 650, which was designed to rent for $3,000 a month—far less than the going price for large computers like the UNIVAC, but nonetheless a lot more than IBM charged for its other office equipment. When it was announced in 1953, those who were backing the project optimistically foresaw a market for 250 machines. They had to convince others in the IBM organization that this figure was not inflated.[7]

As it turned out, businesses snapped up the 650 by the thousands. It became the Model T of computers, and its success helped establish IBM as the dominant computer manufacturer it is today. The idea finally caught on that a private company could manufacture and sell computers—of modest power, and at lower prices than the first monsters—in large quantities. The 650 established the notion of the computer as a machine for business as well as for science, and its success showed that the low estimates of how many computers the world needed were wrong.

Why the inventors and the first commercial manufacturers underestimated the computer's potential market by a wide margin is an interesting question for followers of the computer industry and for historians of modern technology. There is no single cause that accounts for the misperception. Rather, three factors contributed to the erroneous picture of the computer's future: a mistaken feeling that computers were fragile and unreliable; the institutional biases of those who shaped policies toward computer use in the early days; and an almost universal failure, even among the computer pioneers themselves, to understand the very nature of computing (how one got a computer to do work, and just how much work it could do).

It was widely believed that computers were unreliable because their vacuum-tube circuits were so prone to failure. Large numbers of computers would not be built and sold, it was believed, because their unreliability made them totally unsuited for routine use in a small business or factory. (Tubes failed so frequently they were plugged into sockets, to make it easy to replace them. Other electronic components were more reliable and so were soldered in place.) Eckert and Mauchly's ENIAC had 18,000 vacuum tubes. Other electronic computers got by with fewer, but they all had many more than most other electronic equipment of the day. The ENIAC was a room-sized Leviathan whose tubes generated a lot of heat and used great quantities of Philadelphia's electric power. Tube failures were indeed a serious problem, for if even one tube blew out during a computation it might render the whole machine inoperative. Since tubes are most likely to blow out within a few minutes after being switched on, the ENIAC's power was left on all the time, whether it was performing a computation or not.

Howard Aiken was especially wary of computers that used thousands of vacuum tubes as their switching elements. His Mark I, an electromechanical machine using parts taken from standard IBM accounting equipment of the day, was much slower but more rugged than computers that used vacuum tubes. Aiken felt that the higher speeds vacuum tubes offered did not offset their tendency to burn out. He reluctantly designed computers that used vacuum tubes, but he always kept the numbers of tubes to a minimum and used electromechanical relays wherever he could.[8] Not everyone shared Aiken's wariness, but his arguments against using vacuum-tube circuits were taken seriously by many other computer designers, especially those whose own computer projects were shaped by the policies of Aiken's Harvard laboratory.

That leads to the next reason for the low estimates: Scientists controlled the early development of the computer, and they steered postwar computing projects away from machines and applications that might have a mass market. Howard Aiken, John von Neumann, and Douglas Hartree were physicists or mathematicians, members of a scientific elite. For the most part, they were little concerned with the mundane payroll and accounting problems that businesses faced every day. Such problems involved little in the way of higher mathematics, and their solutions contributed little to the advancement of scientific knowledge. Scientists perceived their own place in society as an important one but did not imagine that the world would need many more men like themselves. Because their own needs were satisfied by a few powerful computers, they could not imagine a need for many more such machines. Even at IBM, where commercial applications took precedence, scientists shaped the perceptions of the new invention. In the early 1950s the mathematician John von Neumann was a part-time consultant to the company, where he played no little role in shaping expectations for the new technology.

The perception of a modest and limited future for electronic computing came, most of all, from misunderstandings of its very nature. The pioneers did not really understand how humans would interact with machines that worked at the speed of light, and they were far too modest in their assessments of what their inventions could really do. They felt they had a made a breakthrough in numerical calculating, but they missed seeing that the breakthrough was in fact a much bigger one. Computing turned out to encompass far more than just doing complicated sequences of arithmetic. But just how much more was not apparent until much later, when other people gained familiarity with computers. A few examples of objections raised to computer projects in the early days will make this clear.

When Howard Aiken first proposed building an automatic computer, in 1937, his colleagues at Harvard objected. Such a machine, they said, would lie idle most of the time, because it would soon do all the work required of it. They were clearly thinking of his proposed machine in terms of a piece of experimental apparatus constructed by a physicist; after the experiment is performed and the results gathered, such an apparatus has no further use and is then either stored or dismantled so that the parts can be reused for new experiments. Aiken's proposed "automatic calculating machine," as he called it in 1937, was perceived that way. After he had used it to perform the calculations he wanted it to perform, would the machine be good for anything else? Probably not. No one had built computers before. One could not propose building one just to see what it would look like; a researcher had to demonstrate the need for a machine with which he could solve a specific problem that was otherwise insoluble. Even if he could show that only with the aid of a computer could he solve the problem, that did not necessarily justify its cost.[9]

Later on, when the much faster electronic computers appeared, this argument surfaced again. Mechanical computers had proved their worth, but some felt that electronic computers worked so fast that they would spew out results much faster than human beings could assimilate them. Once again, the expensive computer would lie idle, while its human operators pondered over the results of a few minutes of its activity. Even if enough work was found to keep an electronic computer busy, some felt that the work could not be fed into the machine rapidly enough to keep its internal circuits busy.[10]

Finally, it was noted that humans had to program a computer before it could do any work. Those programs took the form of long lists of arcane symbols punched into strips of paper tape. For the first electronic computers, it was mostly mathematicians who prepared those tapes. If someone wanted to use the computer to solve a problem, he was allotted some time during which he had complete control over the machine; he wrote the program, fed it into the computer, ran it, and took out the results. By the early 1950s, computing installations saw the need for a staff of mathematicians and programmers to assist the person who wanted a problem solved, since few users could be expected to know the details of programming each specific machine. That meant that every computer installation would require the services of skilled mathematicians, and there would never be enough of them to keep more than a few machines busy. R. F. Clippinger discussed this problem at a meeting of the American Mathematical Society in 1950, stating: "In order to operate the modern computing machine for maximum output, a staff of perhaps twenty mathematicians of varying degrees of training are required. There is currently such a shortage of persons trained for this work, that machines are not working full time."[11] Clippinger forecast a need for 2,000 such persons by 1960, implying that there would be a mere 100 computers in operation by then.

These perceptions, which lay behind the widely held belief that computers would never find more than a limited (though important) market in the industrialized world, came mainly from looking at the new invention strictly in the context of what it was replacing: calculating machines and their human operators. That context was what limited the pioneers' vision.

Whenever a new technology is born, few see its ultimate place in society. The inventors of radio did not foresee its use for broadcasting entertainment, sports, and news; they saw it as a telegraph without wires. The early builders of automobiles did not see an age of "automobility"; they saw a "horseless carriage." Likewise, the computer's inventors perceived its role in future society in terms of the functions it was specifically replacing in contemporary society. The predictions that they made about potential applications for the new invention had to come from the context of "computing" that they knew. Though they recognized the electronic computer's novelty, they did not see how it would permit operations fundamentally different from those performed by human computers.

Before there were digital computers, a mathematician solved a complex computational problem by first recasting it into a set of simpler problems, usually involving only the four ordinary operations of arithmetic—addition, subtraction, multiplication, and division. Then he would take this set of more elementary problems to human computers, who would do the arithmetic with the aid of mechanical desk-top calculators. He would supply these persons with the initial input data, books of logarithmic and trigonometric tables, paper on which to record intermediate results, and instructions on how to proceed. Depending on the computer's mathematical skill, the instructions would be more or less detailed. An unskilled computer had to be told, for example, that the product of two negative numbers is a positive number; someone with more mathematical training might need only a general outline of the computation.[12]

The inventors of the first digital computers saw their machines as direct replacements for this system of humans, calculators, tables, pencils and paper, and instructions. We know this because many early experts on automatic computing used the human computing process as the standard against which the new electronic computers were compared. Writers of early textbooks on "automatic computing" started with the time a calculator took to multiply two ten-digit numbers. To that time they added the times for the other operations: writing and copying intermediate results, consulting tables, and keying in input values. Although a skilled operator could multiply two numbers in 10 or 12 seconds, in an 8-hour day he or she could be expected to perform only 400 such operations, each of which required about 72 seconds.[13] The first electronic computers could multiply two ten-digit decimal numbers in about 0.003 second; they could copy and read internally stored numbers even faster. Not only that, they never had to take a coffee break, stop for a meal, or sleep; they could compute as long as their circuits were working.

Right away, these speeds radically altered the context of the arguments that electronic components were too unreliable to be used in more than a few computers. It was true that tubes were unreliable, and that the failure of even one during a calculation might vitiate the results. But the measure of reliability was the number of operations between failures, not the absolute number of hours the machine was in service. In terms of the number of elementary operations it could do before a tube failed, a machine such as the ENIAC turned out to be quite reliable after all. If it could be kept running for even one hour without a tube failure, during that hour it could do more arithmetic than the supposedly more reliable mechanical calculators could do in weeks. Eventually the ENIAC's operators were able to keep it running for more than 20 hours a day, 7 days a week. Computers were reliable enough long before the introduction of the transistor provided a smaller and more rugged alternative to the vacuum tube.

So an electronic computer like the ENIAC could do the equivalent of about 30 million elementary operations in a day—the equivalent of the work of 75,000 humans. By that standard, five or six computers of the ENIAC's speed and size could do the work of 400,000 humans. However, measuring electronic computing power by comparing it with that of humans makes no sense. It is like measuring the output of a steam engine in "horsepower." For a 1- or 2-horsepower engine the comparison is appropriate, but it would be impossible to replace a locomotive with an equivalent number of horses. So it is with computing power. But the human measure was the only one the pioneers knew. Recall that between 1945 and 1950 the ENIAC was the only working electronic computer in the United States. At its public dedication in February 1946, Arthur Burks demonstrated the machine's powers to the press by having it add a number to itself over and over again—an operation that reporters could easily visualize in terms of human abilities. Cables were plugged in, switches set, and a few numbers keyed in. Burks then said to the audience, "I am now going to add 5,000 numbers together," and pushed a button on the machine. The ENIAC took about a second to add the numbers.[14]

Almost from the day the first digital computers began working, they seldom lay idle. As long as they were in working order, they were busy, even long after they had done the computations for which they were built.

As electronic computers were fundamentally different from the human computers they replaced, they were also different from special-purpose pieces of experimental apparatus. The reason was that the computer, unlike other experimental apparatus, was programmable. That is, the computer itself was not just "a machine," but at any moment it was one of an almost infinite number of machines, depending on what its program told it to do. The ENIAC's users programmed it by plugging in cables from one part of the machine to another (an idea borrowed from telephone switchboards). This rewiring essentially changed it into a new machine for each new problem it solved. Other early computers got their instructions from punched strips of paper tape; the holes in the tape set switches in the machine, which amounted to the same kind of rewiring effected by the ENIAC's plugboards. Feeding the computer a new strip of paper tape transformed it into a completely different device that could do something entirely different from what its designers had intended it to do. Howard Aiken designed his Automatic Sequence Controlled Calculator to compute tables of mathematical functions, and that it did reliably for many years. But in between that work it also solved problems in hydrodynamics, nuclear physics, and even economics.[15]

The computer, by virtue of its programmability, is not a machine like a printing press or a player piano—devices which are configured to perform a specific function.[16] By the classical definition, a machine is a set of devices configured to perform a specific function: one employs motors, levers, gears, and wire to print newspapers; another uses motors, levers, gears, and wire to play a prerecorded song. A computer is also made by configuring a set of devices, but its function is not implied by that configuration. It acquires its function only when someone programs it. Before that time it is an abstract machine, one that can do "anything." (It can even be made to print a newspaper or play a tune.) To many people accustomed to the machines of the Industrial Revolution, a machine having such general capabilities seemed absurd, like a toaster that could also sew buttons on a shirt. But the computer was just such a device; it could do many things its designers never anticipated.

The computer pioneers understood the concept of the computer as a general-purpose machine, but only in the narrow sense of its ability to solve a wide range of mathematical problems. Largely because of their institutional backgrounds, they did not anticipate that many of the applications computers would find would require the sorting and retrieval of non-numeric data. Yet outside the scientific and university milieu, especially after 1950, it was just such work in industry and business that underlay the early expansion of the computer industry. Owing to the fact that the first computers did not do business work, the misunderstanding persisted that anything done by a computer was somehow more "mathematical" or precise than that same work, done by other means. Howard Aiken probably never fully understood that a computer could not only be programmed to do different mathematical problems but could also do problems having little to do with mathematics. In 1956 he made the following statement: ". . . if it should ever turn out that the basic logics of a machine designed for the numerical solution of differential equations coincide with the logics of a machine intended to make bills for a department store, I would regard this as the most amazing coincidence that I have ever encountered."[17] But the logical design of modern computers for scientific work in fact coincides with the logical design of computers for business purposes. It is a "coincidence," all right, but one fully intended by today's computer designers.

The question remained whether electronic computers worked too fast for humans to feed work into them. Engineers and computer designers met the problem of imbalance of speeds head-on, by technical advances at both the input and output stages of computing. To feed in programs and data, they developed magnetic tapes and disks instead of tedious plugboard wiring or slow paper tape. For displaying the results of a computation, high-speed line printers, plotters, and video terminals replaced the slow and cumbersome electric typewriters and card punches used by the first machines.

Still, the sheer bulk of the computer's output threatened to inundate the humans who ultimately wanted to use it. But that was not a fatal fault, owing (again) to the computer's programmability. Even if in the course of a computation a machine handles millions of numbers, it need not present them all as its output. The humans who use the computer need only a few numbers, which the computer's program itself can select and deliver. The program may not only direct the machine to solve a problem; it also may tell the machine to select only the "important" part of the answer and suppress the rest.

Ultimately, the spread of the computer beyond physics labs and large government agencies depended on whether people could write programs that would solve different types of problems and that would make efficient use of the high internal speed of electronic circuits. That challenge was not met by simply training and hiring armies of programmers (although sometimes it must have seemed that way). It was met by taking advantage of the computer's ability to store its programs internally. By transforming the programming into an activity that did not require mathematical training, computer designers exploited a property of the machine itself to sidestep the shortage of mathematically trained programmers.

Although the computer pioneers recognized the need for internal program storage, they did not at first see that such a feature would have such a significant effect on the nature of programming. The idea of storing both the program and data in the same internal memory grew from the realization that the high speed at which a computer could do arithmetic made sense only if it got its instructions at an equally high speed. The plugboard method used with the ENIAC got instructions to the machine quickly but made programming awkward and slow for humans. In 1944 Eckert proposed a successor to the ENIAC (eventually called the EDVAC) whose program would be supplied not by plugboards but by instructions stored on a high-speed magnetic disk or drum.

In the summer of 1944, John von Neumann first learned (by chance) of the ENIAC project, and within a few months he had grasped that giant machine's fundamentals—and its deficiencies, which Eckert and Mauchly hoped to remedy with their next computer. Von Neumann then began to develop a general theory of computing that would influence computer design to the present day.[18] In a 1945 report on the progress of the EDVAC he stated clearly the concept of the stored program and how a computer might be organized around it.[19] Von Neumann was not the only one to do that, but it was mainly from his report and others following it that many modern notions of how best to design a computer originated.

For von Neumann, programming a digital computer never seemed to be much of an intellectual challenge; once a problem was stated in mathematical terms, the "programming" was done. The actual writing of the binary codes that got a computer to carry out that program was an activity he called coding, and from his writings it is clear that he regarded the relationship of coding to programming as similar to that of typing to writing. That "coding" would be as difficult as it turned out to be, and that there would emerge a profession devoted to that task, seems not to have occurred to him. That was due in part to von Neumann's tremendous mental abilities and in part to the fact that the problems that interested him (such as long-range weather forecasting and complicated aspects of fluid dynamics[20]) required programs that were short relative to the time the computer took to digest the numbers. Von Neumann and Herman Goldstine developed a method (still used today) of representing programs by flow charts. However, such charts could not be fed directly into a machine. Humans still had to do that, and for those who lacked von Neumann's mental abilities the job remained as difficult as ever.

The intermediate step of casting a problem in the form of a flow chart, whatever its benefits, did not meet the challenge of making it easy for nonspecialists to program a computer. A more enduring method came from reconsidering, once again, the fact that the computer stored its program internally.

In his reports on the EDVAC, von Neumann had noted the fact that the computer could perform arithmetic on (and thus modify) its instructions as if they were data, since both were stored in the same physical device.[21] Therefore, the computer could give itself new orders. Von Neumann saw this as a way of getting a computer with a modest memory capacity to generate the longer sequences of instructions needed to solve complex problems. For von Neumann, that was a way of condensing the code and saving space.

However, von Neumann did not see that the output of a computer program could be, rather than numerical information, another program. That idea seemed preposterous at first, but once implemented it meant that users could write computer programs without having to be skilled mathematicians. Programs could take on forms resembling English and other natural languages. Computers then would translate these programs into long complex sequences of ones and zeroes, which would set their internal switches. One even could program a computer by simply selecting from a "menu" of commands (as at an automated bank teller) or by paddles and buttons (as on a computerized video game). A person need not even be literate to program.

That innovation, the development of computer programs that translated commands simple for humans to learn into commands the computer needs to know, broke through the last barrier to the spread of the new invention.[22] Of course, the widespread use of computers today owes a lot to the technical revolution that has made the circuits so much smaller and cheaper. But today's computers-on-a-chip, like the "giant brains" before them, would never have found a large market had a way not been found to program them. When the low-cost, mass-produced integrated circuits are combined with programming languages and applications packages (such as those for word processing) that are fairly easy for the novice to grasp, all limits to the spread of computing seem to drop away. Predictions of the numbers of computers that will be in operation in the future become meaningless.

What of the computer pioneers' early predictions? They could not foresee the programming developments that would spread computer technology beyond anything imaginable in the 1940s. Today, students with pocket calculators solve the mathematical problems that prompted the pioneers of that era to build the first computers. Furthermore, general-purpose machines are now doing things, such as word processing and game playing, that no one then would have thought appropriate for a computer. The pioneers did recognize that they were creating a new type of machine, a device that could do more than one thing depending on its programmming. It was this understanding that prompted their notion that a computer could do "anything." Paradoxically, the claim was more prophetic than they could ever have known. Its implications have given us the unforeseen computer revolution amid which we are living.

Notes

1. Harold Bergstein, "An Interview with Eckert and Mauchly," *Datamation* 8, no. 4 (1962), pp. 25–30.

2. Simon Lavington, *Early British Computers* (Bedford, Mass.: Digital Press, 1980), p. 104.

3. See, for example, John Wells, The Origins of the Computer Industry: a Case Study in Radical Technological Change, Ph.D. Diss., Yale University, 1978, pp. 93, 96, 119; Robert N. Noyce, "Microelectronics," *Scientific American* 237 (September 1977), p. 64; Edmund C. Berkeley, "Sense and Nonsense about Computers and Their Applications," in proceedings of World Computer Pioneer Conference, Llandudno, Wales, 1970, also in Phillip J. Davis and Reuben Hersh (eds.), *The Mathematical Experience* (New York: Houghton Mifflin, 1981).

4. See, for example, the proceedings of two early conferences: *Symposium on Large-Scale Digital Calculating Machinery*, Annals of Harvard University Computation Laboratory, vol. 16, 1949; *The Moore School Lectures: Theory and Techniques for Design of Electronic Digital Computers*, lectures given at Moore School of Electrical Engineering, University of Pennsylvania, 1946 (Cambridge, Mass.: MIT Press, 1986).

5. Georgia G. Mollenhoff, "John V. Atanasoff, DP Pioneer," *Computerworld* 8, no. 11 (1974), pp. 1, 13.

6. Byron E. Phelps, The Beginnings of Electronic Computation, IBM Corporation Technical Report TR-00.2259, Poughkeepsie, N.Y., 1971, p. 19.

7. Cuthbert C. Hurd, "Early IBM Computers: Edited Testimony," *Annals of the History of Computing* 3 (1981), pp. 163–182.

8. Anthony Oettinger, "Howard Aiken," *Communications ACM* 5 (1962), pp. 298–299, 352.

9. Henry Tropp, "The Effervescent Years: a Retrospective," *IEEE Spectrum* 11 (February 1974), pp. 70–81.

10. For an example of this argument, and a refutation of it, see John von Neumann, *Collected Works*, vol. 5 (Oxford: Pergamon, 1961), pp. 182, 236.

11. R. F. Clippinger, "Mathematical Requirements for the Personnel of a Computing Laboratory," *American Mathematical Monthly* 57 (1950), p. 439; Edmund Berkeley, *Giant Brains, or Machines that Think* (New York: Wiley, 1949), pp. 108–109.

12. Ralph J. Slutz, "Memories of the Bureau of Standards SEAC," in N. Metropolis, J. Howlett, and G. Rota (eds.), *A History of Computing in the Twentieth Century* (New York: Academic, 1980), pp. 471–477.

13. In a typical computing installation of the 1930s, humans worked, with mechanical calculators that could perform the four elementary operations of arithmetic, on decimal numbers having up to ten digits, taking a few seconds per operation. Although the machines were powered by electric motors, the arithmetic itself was always done by mechanical parts—gears, wheels, racks, and levers. The machines were sophisticated and complex, and they were not cheap; good ones cost hundreds of dollars. For a survey of early mechanical calculators and early computers, see Francis J. Murray, *Mathematical Machines*, vol. 1: *Digital Computers* (New York: Columbia University Press, 1961); see also Engineering Research Associates, *High-Speed Computing Devices* (New York: McGraw-Hill, 1950; Cambridge, Mass.: MIT Press, 1984).

14. Quoted in Nancy Stern, *From ENIAC to UNIVAC* (Bedford, Mass.: Digital Press, 1981), p. 87.

15. Oettinger, "Howard Aiken."

16. Abbott Payson Usher, *A History of Mechanical Inventions*, second edition (Cambridge, Mass.: Harvard University Press, 1966), p. 117.

17. Howard Aiken, "The Future of Automatic Computing Machinery," in *Elektronische Rechenmaschinen und Informationsverarbeitung*, proceedings of a symposium, published in *Nachrichtentechnische Fachberichte* no. 4 (Braunschweig: Vieweg, 1956), pp. 32–34.

18. Herman H. Goldstine, *The Computer From Pascal to von Neumann* (Princeton University Press, 1972), p. 182.

19. Von Neumann's "First Draft of a Report on the EDVAC" was circulated in typescript for many years. It was not meant to be published, but it nonetheless had an influence on nearly every subsequent computer design. The complete text has been published for the first time as an appendix to Nancy's Stern's *From ENIAC to UNIVAC* (note 14).

20. Von Neumann, *Collected Works*, vol. 5, pp. 182, 236.

21. Martin Campbell-Kelley, "Programming the EDSAC," *Annals of the History of Computing* 2 (1980), p. 15.

22. For a discussion of the concept of high-level programming languages and how they evolved, see H. Wexelblatt (ed.), *History of Programming Languages* (New York: Academic, 1981), especially the papers on FORTRAN, BASIC, and ALGOL.

10

Dazzling the Multitude: Imagining the Electric Light as a Communications Medium

Carolyn Marvin

Marshall McLuhan, a popular media prophet of the 1960s, believed that the history of Western culture should be rewritten so as to cast successive new technologies of communication in the role of the great levers that moved it. Not the message of communication, McLuhan argued, but the medium—the structural characteristics of the techniques and machines of information storage, retrieval, and transmission—had a semiotic eloquence that overshadowed the particular details of the content. The medium, McLuhan declared, "shapes and controls the scale and form of human association and action."[1] McLuhan's account of cultural evolution in the West has found little favor among historians, but his appreciation of the relationship between technology and culture and his colorful efforts to spotlight that relationship helped focus the problem for others.[2] That relationship is now a staple concern of scholarship in the history of technology.

McLuhan's definition of an information medium was very broad. He was fond of insisting that even the electric light is an information medium. That this example was intended to shock McLuhan's readers is a measure of the historical distance we have traveled, for this claim would not have seemed nearly as strange in the popular or the scientific culture of late-nineteenth-century Europe and the United States (albeit in a somewhat different sense than McLuhan intended). This chapter is an attempt to reconstruct that forgotten dimension of the social history of electricity by tracing some early contributions of the electric light to the complex transformation of communication that began with the telegraph, proceeded through the electronic mass media, and continues at the present moment in computing technology.

In the late nineteenth century, some prophets of the future imagined great banks of electric lights spelling out letters and pictures to astonish passersby, or mammoth searchlights projecting stenciled messages and images on the clouds for the pleasure and information of all in the surrounding countryside. What was imagined was also occasionally attempted.

The communicative utility of the electric light did manifest itself in some areas; traffic lights, movie marquees, and neon signs still testify to this. However, the principal legacy of the electric light to modern mass communications was something different, something not foreseen: The electric light (particularly the incandescent light) helped transform public spectacles from traditional outdoor community gatherings lit by candles, bonfires, or oil lamps into the glittering indoor mass-media spectacles familiar to us.

Throughout the United States and Western Europe the public introduction of the electric light excited interest and curiosity. In 1886 the editors of the *Electrical Review* recalled the first appearance of electric lighting in New York shop windows as follows:

> It was looked upon as a mere experiment, the continuation of which would soon prove more trouble than it was worth, and the neighboring stores took no stock in it. Soon, however, it was discovered that it was attracting the attention of customers and the general public to such an extent that its users were compelled to enlarge the stock. Owing to the brilliancy of the light pedestrians could walk by stores of the same character lighted by gas without even seeing them, so attractive was the brilliant illumination further along. They clustered and fluttered about it as moths do about an oil lamp. That settled it. The neighboring stores must have it, and the inquiry and demand for the light spread apace until now, when, as soon as the electric light appears in one part of a locality in an American city it spreads from store to store and from street to street.[3]

The scientists, engineers, and entrepreneurs upon whose comments and predictions this study is mainly based framed their expectations of electricity in terms of the only technological revolution familiar to them: that brought about by steam power. Many thought the impact of electricity would exceed even that great leap, completing its promise and salvaging its disappointments. Abundant, easily distributed, versatile electricity would reverse the centralization of production in factories, lead to the rise of clean cottage industries, unify the home and the workplace, and lower the divorce rate. By decentralizing the population, it would make the cities green with parks and gardens. Some nineteenth-century observers believed that the concentration of labor around steam-powered urban factories had made the visible differences between the working life of the city and the leisure life of the country starker than ever. They looked to electrical manufacturing and transportation to create a homogeneous landscape that would heal the breach between classes which steam had exacerbated. They also expected electricity to democratize luxury and eliminate conflict based on competition for scarce resources by producing goods cheaply and abundantly. Not all predictions were so optimistic, but a great many were concerned with these possibilities.

As a purveyor of message content, the electric light was not much of a novelty; it simply extended the earlier uses of signal lights to transmit warnings and news. However, the illumination of public places by electric filament lamps seemed much more dramatic, more colorful, more elaborate, and more versatile than the village bonfire (which predated the Middle Ages), the floating-wick oil lamp of the eighteenth-century garden fête, or the carbon arc lamp. Like its traditional predecessors, electric lighting conveyed the message that the occasion of its use was exciting and vivid. Unlike their traditional predecessors, electrically lighted events were often commercially sponsored and organized.

The most striking electric-light spectacles were the great industrial exhibitions of the late nineteenth century. The Chicago World's Fair and Columbian Exposition of 1893 was one of the most splendid and one of the most self-consciously electric. By the time of this fair the promise of electricity occupied a place near the center of popular enthusiasm, as expressed in the popular and trade press. The fair had an advanced telephone and telegraph system for internal and external communication; electrically powered boats, rail cars, and moving sidewalks; great electric motors that operated machinery; exhibits of the latest electrical inventions, and 90,000 electric lights mounted on buildings, walkways, and illuminated displays.[4] One of the most popular attractions was the 82-foot-tall Edison Tower of Light, a pedestal covered with multicolored incandescent bulbs.[5] Another was the evening show at the Court of Honor, where jets of water from electrically powered fountains and flashing electric lights combined in fanciful patterns. In the lights of the Court of Honor one observer saw "great flowers, sheaves of wheat, fences of gold, showers of rubies, pearls and amethysts." Another sensed a nearly evangelical power in the nighttime illumination of the fair:

> Under the cornices of the great buildings, and around the water's edge, ran the spark that in an instant circled the Court with beads of fire. The gleaming lights outlined the porticoes and roofs in constellations, studded the lofty domes with fiery, falling drops, pinned the darkened sky to the fairy white city, and fastened the city's base to the black lagoon with nails of gold. And now, like great white suns in this firmament of yellow stars, the search lights pierced the gloom with polished lances, and made silvern

paths as bright and straight as Jacob's ladder, sloping to the stars or shooting the beams in level lines across the darkness, effulgent milky ways were formed or again, turned upward to the zenith, the white stream flowed toward heaven until it seemed the holy light from the rapt gaze of a saint or Faith's white, pointing finger.[6]

The spectacular effect of electric light in public spaces was a subject of interested discussion long before such lighting became a staple of fairs and expositions. In 1884 *Electrical World* claimed that only the electric light was "considered worthy or suitable to illuminate conspicuous and beautiful public buildings," and that electric lighting was the only form of decoration being considered for the Statue of Liberty. *Electrical World* reported several plans to light the statue, including one to project vertical beams of light upward from the torch to stand as "a pillar of fire by night." A second plan was to place lights "like jewels around the diadem," and a third was to place them at the foot of the loggia to illuminate the entire statue, so that "the illumined face of Liberty [would] shine out upon the dark waste of waters and the incoming Atlantic voyagers."[7] An imaginative journalist had proposed that the statue should hold aloft the world's largest electric light "to illumine the lower bay and even to make Coney Island, which with its myriads of lights glistens on a summer's night like a huge diamond, pale and insignificant, and like the evening star when the moon is in full form."

In a world where electric lights are prosaically utilitarian and unremarkably plentiful, such descriptions may be understood as a reaction to something novel in the experience of enthusiastic observers: the introduction on a new scale of the grand illusion, the effect that also most clearly defines success in modern mass media. Accustomed as we are to electric lights and to more elaborate illusions, it is difficult for us to imagine the original impact of the electric-light spectacle. By reilluminating nature, the electric light offered some observers a way of rediscovering it. It offered others a novel means to manufacture and sell what they described as the improved experience of nature.

New York Harbor was the setting for a number of electric-light shows in the 1880s. From the shore of Staten Island, Manhattan at night was said to look like a "fairyland" during one such show, with "a thousand electric lights dancing from out a sea of inky gloom, with here and there a cross, and there a crown, near which fireflies of huge dimensions start here and there with phosphor fires aglow, the streets ashimmer with silver, with calcined towers lumined against the unfathomable gloom beyond."[8] On Staten Island in 1886 a colored fountain inspired applause from opening-night spectators as it was put through its paces: "At one moment it was crystal, the next roseate, then successively green, blue, purple, gold, and from time to time the tints would blend, harmonize, and contrast with new charms at every change. . . . Far out in the bay it could be seen, looking like a gigantic opal, illuminated by its internal fires."[9] After a visit to the United States, William Preece, soon to be Chief Engineer of the British Post Office, and an attentive observer of American ingenuity, described for lecture audiences the commercial potential in the splendid spectacle of Brooklyn Bridge lit by 82 arc lamps: "It is so beautiful, in a scenic sense, that one of the enterprising ferry companies contemplates having nightly excursions during the summer season, which it is intended to advertise as the 'Theater of New York Harbor by Electric Light; price of admission, 10 cents.' "[10]

Outdoors the electric lamp was safer, cheaper, and more versatile than carbon arc lighting. Indoors it eliminated the disagreeable fumes of low-candlepower gas lighting and the intense glare and uncertain safety of arc lighting. Gradually, some outdoor events began to move indoors to the smaller settings for which incandescent lights were ideally suited. Fancy lights were laid on for sumptuous balls, receptions, and banquets, and entertainers appropriated the electric light as a performance prop.

In 1884 the Electric Girl Lighting Company offered to supply "illuminated girls" for indoor occasions. Young women hired to perform the duties of hostesses and serving girls while decked out in filament lamps were advertised to prospective customers as "girls of fifty-candle power each in quantities to suit householders."[11] The women were fed and clothed by the company, and customers were "permitted to select at the company's warehouse whatever style of girl may please their fancy." In Kansas City, employees of the Missouri and Kansas Telephone Company organized a public entertainment in 1885 in Merchants' Exchange Hall which was graced by "an electric girl, placed on exhibition" along with a model switchboard and telephone exchange.[12] "Electric girls" embodied both the personal servant of a passing age, a potent symbol of social status, and the electric light as ornamental object, a dazzling and opulent display of social status in a new age. In time, impersonal electricity would help banish most personal servants, and would make electric lighting essential and functional for all classes instead of a badge of conspicuous consumption for one. Indeed, electric girls were already transitional, since they were not traditional family servants but were hired for the occasion.

Entertainers also began to use electric lights to adorn their bodies in public performance. Before an astonished audience in Sheldon,

Iowa, in 1891, Miss Ethyl J. Davis of the Ladies Band Concert and Broom Brigade rendered a tableau of the Statue of Liberty, which the local newspaper reported in detail:

> Miss Davis stood on an elevated pedestal, her upraised right hand grasping a torch, capped with a cluster of lamps, which alone would have been sufficient to illuminate the entire room. A crown with a cluster of lamps, and covered with jewels, and her robes completely covered with incandescent lamps of various sizes and colors completed the costume. The lights in the hall were turned down and almost total darkness prevailed. As the contact was made bringing the electric lamps into circuit, the entire hall was illuminated with a flood of light which almost blinded the spectators, and Miss Davis, standing revealed in the glaring light, certainly presented a picture of unparalleled brilliance and beauty.[13]

"The Greatest Event in the history of Brookings, South Dakota" was the description given by a local newspaper to a Merchants' Carnival held in 1890 at the Brookings opera house at which various industrial enterprises were represented by appropriately costumed ladies. One of them was Mrs. E. E. Gaylord, wife of the manager and electrician of the Brookings Electric Light Company. To represent that flourishing branch of commerce, Mrs. Gaylord wore

> a crown of incandescent lamps and her dress was decorated with the same ornaments. The lamps were all properly connected, the wires terminating in the heels of the shoes. On the floor of the stage were two small copper plates connected to a small dynamo. When Mrs. Gaylord touched the plates the 21 lamps of her crown, banner and costume instantly flashed up and she stood clad in "nature's resplendent robes without whose vesting beauty all were wrapt in unessential gloom."[14]

In 1893 George W. Patterson of Chicago created a special novelty act of "electrical spectacular effects" with lighted Indian clubs. By swinging these clubs in a darkened room, he created the illusion of circles and other designs of solid light. Describing his act in 1899, the *Western Electrician* detailed a "striking feature" of the entertainment, the "electrical storm":

> . . . beginning with distant heat lightning, gradually increasing to the fiercest of chain or "zigzag" lightning, with corresponding graduation of thunder, the latter being produced in the usual manner by a "thundersheet" of iron. . . . The effect is very startling, especially as it is accompanied by the fiercest thunder, the sound of dashing rain and by Mr. Patterson's voice laughing and singing "The Lightning King" through a megaphone. The "Lightning King" is followed by the latter part of "Anchored," in which a

Figure 1
Mrs. E. E. Gaylord representing electrical enterprise at the Merchants' Carnival, Brookings, South Dakota, 1890.

perfect double rainbow gradually appears and is dissolved by a water rheostat, by sending the rays of a single-loop-filament incandescent lamp through a prism. The colors come out beautifully.[15]

The familiar dimensions of bodily experience have always provided a reference point for exploring the significance and utility of new and unfamiliar technologies. The social uncertainties created by the introduction of novel, various, and intellectually mysterious technologies are reduced and appropriated for a variety of purposes by recasting them in this familiar idiom. Electric lights even appeared in jewelry. A New York exhibition in 1885 of "flash" jewelry from Paris included hatpins and brooches studded with tiny glittering electric lights.[16] The use of electric lights and effects in the realm of fashion was described in detail by *Western Electrician* in 1891:

Electric jewelry usually takes the form of pins, which are made in various designs. One such trifle copies a daisy, and has an electric spark flashing from the center. Another is a model of a lantern in emerald glass, while a death's head in gold, with a ray gleaming from each eye, is a third. The wearing of electric jewelry necessitates the carrying about of an accumulator which resembles a spirit flash, and is generally stowed in a waistcoat pocket. Brooches are made occasionally for ladies' wear, but as women have no available pockets, a difficulty arises with regard to the battery.[17]

The impulse to dazzle an audience with electrical effects found expression not only in entertainments in which the spectacle itself was the message but also in the construction from electric lights of illuminated letters and simple figural representations. Antecedents for texts of light go back at least to the illuminated sign in St. James's Park, London, in 1814, in which a star and the words, "the Peace of Europe," were created from more than 1,300 spout-wick oil lamps attached to iron frames to celebrate (prematurely, as it turned out) the end of the Napoleonic Wars.[18] The electric filament lamp made such achievements simpler and inspired still more ambitious ones.

A wedding in Atlanta in 1899 featured illuminated textual decorations. The groom, an electrical contractor, had draped 200 incandescent lights from one side of the church to the other. Directly above the altar hung a wedding bell fashioned of foliage and 100 colored lights. Further details were reported:

An arc light suspended from the interior of the bell represented the clapper. To the right of the bell a letter N, the initial letter of the name of the bride, formed of white incandescents, set in pink flowers, was supported, on invisible wires. A letter L for the groom was on the left. . . . . As the bride with her brother entered the church by one aisle and the groom with his best man entered by the other the letters N and L flashed into view and sparkled with great splendor. A murmur of admiration arose from the auditorium at the superb effect.[19]

As the minister pronounced the pair man and wife, "a single letter L in pink incandescents appeared on the bell and burned with a soft brilliance."

Texts and figures constructed from electric lights became popular for advertising. A common device was the "sky sign," which spelled out the name of a firm or promotional slogan or outlined an image against the blank wall of a building. An especially ambitious sign was erected in 1892 on the side of the 22-story Flatiron Building at the confluence of Broadway and Fifth Avenue with 23rd Street in New York. This sign consisted of 1,500 incandescent lights, white, red, blue,

Figure 2
Electrically illuminated sky signs, 1892.

and green, arranged in seven sentences of letters 3–6 feet high. Each of the seven sentences lit up in succession in a different color from dusk until 11 o'clock each night and brought "to the attention of a sweltering public the fact that Coney Island . . . is swept by ocean breezes," reported *Western Electrician*.[20] "So long as the changes are being run," the account continued, "the public is attracted and stands watching the sign, but as soon as the whole seven sentences are lighted and allowed to remain so, the people move on their way and the crowd disperses. This illuminated sign is not only a commercial success, but when all the lamps are lighted is really a magnificent sight. Its splendor is visible from away up town."

To attract crowds on the streets of large cities, advertisers often used incandescent lights and magic lanterns in combination. On "magic lantern avenues" in Paris, commercial messages were projected on shop windows high above the street. In London "diverting and artistic" displays of the same kind were set up in the Strand and in Leicester Square as early as 1890.[21] Magic lanterns also projected "living photograph" advertisements on billboards, on pavements, and even on Nelson's Column (until the Office of Works prohibited this "desecration" in 1895).[22] In Edinburgh, an electric signboard in front of the Empire Palace Theatre flashed out in 130 colored lights the words *Empire* and

Palace alternately, so that "one word [appeared] in the position previously occupied by the other."[23]

Patriotic and political events were also occasions for electric-light messages. In 1891 the national convention of the Grand Army of the Republic at Detroit was illuminated by an outdoor electrical design, the principal feature of which was a badge of light 48 feet high and 16 feet wide, inscribed with an eagle, a flag, and a cannon. "The words 'Hail, Victorious Army' were shown in letters seven feet high, 600 16-candle power lamps being required for this alone. In addition the G. A. R. monogram was shown in 12-foot letters of red, white, and blue lights. This was visible, it was said, for ten miles down the Detroit river. There was also an anchor, representing the navy, and a horse's head, standing for the cavalry. These took about 100 lights each."[24] In addition, 2,000 lights mounted on and surrounding city hall created a glow that could be seen for 5 miles in every direction.

In 1897 the city of Berlin celebrated the ninetieth birthday of Emperor Wilhelm with a "grand illumination." Multicolored lights arranged in the initials of the emperor and the empress and the significant dates of their reign were strung across public buildings and private houses.[25]

In the United States, the intense feeling kindled by Admiral Dewey's return from the Philippines in 1899 was expressed in dramatic displays of public support all over the country. In New York, the scene of the admiral's triumphant homecoming, an enormous "Welcome Dewey" sign in lighted letters 36 feet high was stretched 370 feet across the Brooklyn Bridge. The letter W alone consisted of 1,000 lights.[26] Chicagoans mounted an electric light picture of Dewey's flagship, the *Olympia*, on scaffolding at the corner of State Street and Madison. "The ship itself was outlined by 720 eight-candle power lamps, 200 red-bulb and 150 blue-bulb lamps being used. A 10,000 candle power, 35-ampere searchlight was placed in the pilothouse of the ship. Portraits of Dewey and McKinley were outlined by incandescent lamps."[27]

With the development of a national telegraphic network in the United States during the second half of the nineteenth century, it became customary for crowds to gather in the streets of many towns and cities on presidential election nights, partly for entertainment and partly to keep track of late returns, which were posted outside local newspaper offices where the wires ran. In 1896, an estimated 250,000 New Yorkers celebrated McKinley's victory in the streets as incoming electoral tallies were projected by calcium light on buildings all across the city. Returns were projected on the outside of the *New York Times* building beneath the slogan "All the News That's Fit to Print," which was spelled out in electric lights next to a magic-lantern portrait of the president-elect.

In that news and entertainment were presented dramatically and in rapid sequence, this scene in Times Square can be said to have prefigured the electronic broadcasting of the twentieth century. The crowds swirling in the street were a prototype mass audience. *Harper's Weekly* took this rather lightly, although it did note that the crowds were "entertained as well as instructed."[28]

Upon such experiments and spectacles future schemes of communication by electric light were erected with great imaginative flourish. To a twentieth-century-accustomed vision, the most fanciful were the proposals to inscribe the night skies with powerful beams of light that could be seen by all inhabitants of the surrounding countryside. This proposal, which appeared in many variations, was a plausible and promising technological extrapolation from existing achievements. It extended both the familiar principle of magic-lantern projection and existing applications of electric-light technology, including some newly hatched attempts to improve the reliability and safety of shipping.

The fact that vessels sailing the coast could often see the locations of towns from the reflections of their night lights off overhanging clouds inspired a series of experiments in which brilliant Morse-code flashes were projected overhead from naval vesssels. In one experiment the flashes were decipherable at a distance of 60 miles.[29] In another, an astonished crowd filled the streets in the vicinity of the Siemens-Halske factory in Berlin, where a light-projecting apparatus strong enough to illuminate handwriting at the distance of a mile was aimed at the sky. With the help of a large mirror, signals placed in front of the light were "repeated, of course on a gigantic scale, on the clouds."[30]

The prospect of illuminated messages on the slate of the heavens fascinated experts and laymen alike. "Imagine the effect," said the *Electrical Review*, "if a million people saw in gigantic characters across the clouds such words as 'BEWARE OF PROTECTION' and 'FREE TRADE LEADS TO H—L!' The writing could be made to appear in letters of a fiery color."[31] According to one electrical expert, "You could have dissolving figures on the clouds, giants fighting each other in the sky, for instance, or put up election figures that can be read *twenty miles away*."[32]

Since such projects were usually undertaken for commercial ends, the popular term for celestial projection was "advertising on the clouds." In 1889 an American inventor claimed to be negotiating with several firms that wished to "display their cards" on the sky.[33] A few years later in England, an experimenter was said to have succeeded in producing the letters BUF upon the clouds, although his target was apparently too small to accommodate the rest of the message: FALO BILL'S WILD WEST.[34] Advertising on the clouds, explained E. H. Johnson, one of Edison's close associates, was "simply a stereopticon on a large scale" that required a light sufficiently powerful to project on the firmament and a method for focusing diffuse light on a "cloud canvas" constantly shifting its distance from the earth.[35] "Even portraits are said to have been 'placed' on the clouds," stated one account, "though the report does not say how great the resemblance was."[36]

While conducting experiments with a large searchlight atop Mount Washington in New Hampshire over a period of months, a General Electric engineer, one L. Rogers, received letters from viewers as far away as 140 miles. He marveled to think that "hundreds of thousands of eyes were centered on that one single spot, waiting for the flash or wink of the 'great luminous eye.' "[37] Reflecting on the Mount Washington experiment a year later, Amos Dolbear, one of the inventors of telephony, imagined the day when great stencil sheets of tin and iron would be prepared for projecting on the clouds an "advertising-sheet" with letters more than 100 feet long that could be read a mile or more away, and when weather forecasts would be "given by a series of flashes of long and short duration, constituting a code of signals."[38] The eventual outcome of Rogers's experiments in casting "legible lines" on the clouds was a huge, electrically powered magic lantern erected in 1892 atop Joseph Pulitzer's *World* building, then the tallest building in New York. It had an illumination of some 1,500,000 candles, and it weighed well over 3,000 pounds. An 8-inch lens projected stencil-plate slides of figures, words, and advertisements upon the clouds or, on clear nights, on the buildings nearby.

Objections to the vulgarity of advertising on the sky were common. In 1892, *Answers*, the penny-journal flagship of the British publishing empire headed by Lord Northcliffe, called the possibility of sky signs the "newest horror" in an article that went on to say:

> You will be able to advertise your wares in letters one hundred feet long on the skies, so that they will be visible over a dozen countries. As if this truly awful prospect were not enough, we are told that these sky-signs can be made luminous, so that they will blaze away all night! A poet, in one of his rhapsodies, said that he would like to snatch a burning pine from its Norway mountains and write with it the name of "Agnes" in letters of fire on the skies.
>
> But he would probably not have cared to adorn the firmament with a blazing description of somebody's patent trouser-stretcher, or a glowing picture, as large as Bedford Square, of a lady viewing the latest thing in corsets.[39]

Another British journal decried "celestial advertising" as follows:

> . . . the clouds are to be turned into hideous and gigantic hoardings. This awful invention deprives us of the last open space in the world on which the weary eye might rest in peace without being agonized by the glaring monstrosities wherewith the modern tradesman seeks to commend his wares.[40]

If the sky was one surface upon which to project messages for the millions, the moon was another. *Science Siftings* reported in 1895 that an American named Hawkins planned to send a flashlight message from London to New York by way of the moon. Using the only satellite available in 1895, Mr. Hawkins had conceived the intellectual principles of satellite relay. The value of his plan, he announced, lay in covering long distances, "but electricity would be required for local distribution from the receiving stations. If a flash of sufficient strength could be thrown upon the moon to be visible to the naked eye, every man, woman and child in all the world within its range could read its messages, as the code is simple and can be quickly committed to memory."[41]

Signaling schemes to strike up a wireless conversation with extraterrestrial beings received wide publicity. "With our powerful searchlights it would be possible to communicate with the planet Mars," Amos Dolbear wrote in his regular science column in *Cosmopolitan* in 1892, "if it should chance to be peopled with intelligences as well equipped with lights and telescopes as we are."[42] Others proposed semaphoric arrangements of giant lights flashing in Morse-code sequence. *Live Wire*, a dime monthly whose title bore witness to the popular association between excitement, novelty, and electricity, reported that Sir Francis Galton had proposed the construction of 75-by-45-foot heliographic mirrors to flash a regularly pulsing ray of sunlight to Mars. Charles Cross thought the beam of a powerful electric light might be gathered and concentrated at a single point by huge parabolic reflectors.[43] There was a proposal that "incandescent lights be strung over the sides of the Great Pyramid, and thus it be made a great square of light." According to *Live Wire*, "When it was pointed out how inadequate this would be, the proposer replied by saying, 'Then illuminate all the pyramids.' "[44] The French science-fiction novelist Camille Flammarion suggested grouping immensely powerful lights in the pattern of the Big Dipper to catch the eyes of extraterrestrial observers. The lights could be placed at Bordeaux, Marseilles, Strasbourg, Paris, Amsterdam, Copenhagen, and Stockholm. "But no one has yet been found to build seven lights each of about three billion candlepower," explained *Live Wire*. A suggestion to work out a geometrical problem in lights for the amusement of galactically remote viewers was impractical, *Live Wire* concluded, because the lines of every figure would have to be at least 50 miles wide and "made of solid light." E. W. Maunder, Great Britain's Assistant Astronomer Royal, speculated that "if ten million arc-lights, each of one hundred thousand candle-power were set up on Mars, we might see a dot."[45]

A repeated assumption in imaginative portrayals of mass audiences of the future was that such audiences properly belonged outdoors, and that twentieth-century media would provide regular occasions for outdoor assemblies. In these speculations, familiar nineteenth-century images of spectators clustered about terrestrial illumination were expanded to the grander scale deemed suitable for illuminated celestial displays. As it has turned out, mass audiences do not collect outdoors to view electric-light messages in the night sky. However, other elements of these speculations and of the early illuminated gatherings prefigure the most familiar of modern public spectacles: television broadcast entertainment. The so-called television special and other broadcasting genres still use dramatic arrangements of brightly colored lights to create visual excitement.

Television's inheritance from the electric light is technological as well as social. Though poorly understood at the time, the original electronic effect, the Edison effect, was created in an electric lamp whose vacuum bulb was the forerunner of the tube that would soon become the principal vehicle of broadcasting. The development of electronics out of this puzzle in a light bulb eventually helped make many face-to-face public gatherings nearly superfluous as families retreated indoors to watch on their private television sets the descendants of the public spectacles that once would have entertained them in the town square. It is not uncommon for technological innovations intended to streamline, simplify, or enhance familiar social routines to so reorganize them that they become new events. Incandescent lights not only inspired new outdoor gatherings, such as night baseball games; they also transformed many large outdoor community gatherings into indoor private ones the size of a single family.

Because communication at a distance was actually implemented in other forms, our cultural memories no longer include predictions made in nineteenth-century voices that the media of the future might be messages of light splashed across the firmament by searchlights or great banks of flashing incandescent lamps. Our amnesia is testimony to the tendency to read history backward from the present—to see it as the process by which our ancestors looked for and gradually discovered us rather than as a succession of self-contained accounts of a moral order, each with its own focused concerns and its own peculiar sense of inhabiting the crucial if not the final stages of human history. If every present attempts to colonize the past with its own spirit, it also appropriates the future with equal enthusiasm. The nineteenth-century conviction that important twentieth-century media would look like nothing so much as the nineteenth-century electric light writ large betrays the companion tendencies to read the past as a less glamorous version, and the future as a more glamorous version, of the present. In the last analysis, the utility of social prediction stands least of all on its accuracy as a pointer to the future. It stands much more on what it communicates about the perceptions, values, and imaginative reactions of societies to changes which they not only must devise ways of coping with and adjusting to but which they in large part also shape.

Notes

1. Marshall McLuhan, *Understanding Media: The Extensions of Man* (New York: McGraw-Hill, 1964), p. 9.

2. Elizabeth Eisenstein, for example, has recently acknowledged her debt to McLuhan while repudiating many of his conclusions about the impact of printing. See *The Printing Press as an Agent of Change: Communications and Cultural Transformations in Early-Modern Europe* (Cambridge University Press, 1979), volume 1, pp. x–xvii, 16–17, 40–41.

3. "Turning Off the Gas in Paris," *Electrical Review*, September 18, 1886, p. 4.

4. Trumbull White and William Igelheart, *The World's Columbian Exposition, Chicago, 1893* (Philadelphia: International, 1893), p. 302.

5. Ibid., pp. 322–323. See also Ben C. Truman, *History of the World's Fair, Being a Complete Description of the World's Columbian Exposition From Its Inception* (Chicago: Cram Standard, 1893), pp. 358–359.

6. Rossiter Johnson (ed.), *A History of the World's Columbian Exposition* (New York: Appleton, 1897), volume 1, p. 510.

7. "Lighting the Statue of Liberty," *Electrical World*, April 26, 1884, p. 136.

8. *Electrical Review*, September 1, 1888, p. 4.

9. *Electrical Review*, July 10, 1886, p. 9.

10. Remarks by William Preece in a speech before the Society of Arts in London, quoted in "The Light That Will Extinguish Gas," *Electrical Review*, February 7, 1885, p. 2.

11. "The Use of Illuminated Girls," *Electrical World*, May 10, 1884, p. 151.

12. *Electrical Review*, February 7, 1885, p. 4.

13. "Electricity in Iowa," *Western Electrician*, June 27, 1891, p. 367.

14. "The Representative of the Electric Light," *Western Electrician*, April 12, 1890, p. 210.

15. "Electrical Spectacular Effects," *Western Electrician*, April 8, 1899, p. 196.

16. "Trouve's Jewelry," *Electrical Review*, June 27, 1885, p. 2.

17. *Western Electrician*, November 7, 1891, p. 1.

18. William T. O'Dea, *The Social History of Lighting* (London: Routledge and Kegan Paul, 1958), p. 178.

19. "A Wedding with Electrical Accessories," *Western Electrician*, December 30, 1899, p. 381.

20. "Electrically Illuminated Signs," *Western Electrician*, December 30, 1899, p. 381.

21. "Smart Advertising Booms," *Answers*, August 2, 1890, p. 150.

22. *Electrical Engineer*, April 26, 1895, p. 18.

23. "Electrical Advertising," *Electrical Engineer*, January 4, 1895, p. 30.

24. "Electrical Decorations at Detroit," *Western Electrician*, September 12, 1891, p. 153.

25. "Electricity and the Birthday of the German Emperor," *Electrical Review*, May 14, 1887, p. 3.

26. "Electrical Decorations in New York," *Western Electrician*, October 14, 1899, p. 217.

27. "Electrical Illumination at the Chicago Festival," *Western Electrician*, October 14, 1899, p. 217.

28. "Election Night in New York," *Harper's Weekly*, November 14, 1896, p. 1122.

29. "Cloud Telegraphy," *Electrical Review*, May 5, 1888, p. 5. Reprinted from *Youth's Companion*, no date. The two vessels were the *Orion* and the *Espoir* of the British Navy. "The *Orion*, having thrown upon the clouds a regular messsage by means of successful flashes, this message was read and understood on board the *Espoir*."

30. "The Electric Light as a Military Signal," *Scientific American*, October 30, 1875, p. 281.

31. *Electrical Review*, October 6, 1888, p. 4.

32. "Advertising in the Clouds: Its Practicability," *Electrical World*, December 31, 1892, p. 427.

33. *Electrical Review*, December 21, 1889, p. 4.

34. "Even the Clouds Don't Escape Him," *Electrical World*, November 26, 1892, p. 335.

35. "Advertising in the Clouds" (note 32).

36. "Even the Clouds Don't Escape Him" (note 34).

37. "Advertising on the Clouds," *Invention*, February 17, 1894, pp. 150–151.

38. Amos E. Dolbear, "The Electric Searchlight," *Cosmopolitan*, December 1893, p. 254.

39. "The Newest Horror," *Answers*, July 16, 1892, p. 129.

40. "Even the Clouds Don't Escape Him."

41. "A Message From the Moon," *Science Siftings*, November 16, 1895, p. 77.

42. Dolbear, "The Electric Searchlight" (note 38).

43. Arthur Bennington, "Some of the Plans of Science to Communicate with Mars, 40,000,000 Miles Away in the Depths of Infinite Space," *Live Wire*, February 1908, p. 6.

44. Ibid.

45. Ibid.

Epilogue

Joseph J. Corn

The popular prophecies and visions of the future discussed in the above chapters reveal not only the fallibility of prediction but also a history of human hopes and fears. Even the most flawed or utopian dream offers insights into how people thought about their world, about social change, about themselves, and about their technology. Visions of the future always reflect the experience of the moment as well as memories of the past. They are imaginative constructs that have more to say about the times in which they were made than about the real future, which is, ultimately, unknowable.

The authors have raised three major issues about these visions of the future. The first relates to the fact that many predictions have been erroneous, exaggerated, or wildly utopian. What explains the hyperbole of so many American predictions? What values, social factors, or aspects of the technologies themselves have influenced the extravagant way the future has been perceived? The second issue concerns the popularity of such buoyant beliefs. Just who believed the predictions, made in newspapers, popular periodicals, and books, of a world transformed by technology? The third issue has to do with the consequences of such beliefs. How has the penchant for extravagant prediction influenced American society? How, for example, has this kind of thinking affected technological development?[1]

Let me propose three common fallacies that have contributed to the extravagant and often utopian tone of technological prediction: the fallacy of total revolution, the fallacy of social continuity, and the fallacy of the technological fix.

Those who in the 1940s and the 1950s envisioned nuclear energy wholly supplanting other sources of power illustrate the fallacy of total revolution. They prophesied atomic-powered automobiles, airplanes, trains, and even washing machines, and a day when electricity would be inexpensive if not altogether free. There would be no need for oil, coal, water power, or any other non-nuclear fuel in the halcyon atomic age. So complete would be this revolution that universal abundance, ample leisure, and social harmony would flow from the splitting of the atom. In short, the adoption of nuclear technology would quickly bring about a new age.

Nuclear enthusiasts failed to foresee the problems of developing nuclear energy. They vastly underrated safety hazards and overrated the ease with which reactors might be made smaller and lighter. Most important, they failed to grasp the fact that new innovations seldom become universal overnight. The history of science and technology shows that only through long use are complex devices sufficiently improved to find widespread application, and that through such use problems not envisioned at the outset—such as what to do with all the atomic waste generated by nuclear technologies—often emerge to further retard acceptance of the technology. Prophets of a total atomic revolution also failed to understand that the introduction of a radically new technology often stimulates significant improvements to competing technologies (in this case, conventionally fueled power plants).[2] Ironically, one of the new technologies that permitted more economical consumption of coal, oil, and gas in power plants has been the microprocessor, yet the development of the transistors and integrated circuits on which it depends was itself the result of an "unforeseen revolution," as Paul Ceruzzi has demonstrated in his chapter. Thus, the total revolution envisioned by advocates of nuclear power was in part thwarted by a failure of vision regarding another realm of technological development.

The fallacy of social continuity represents the other side of the coin. Those who succumb to the fallacy of total revolution envision everything changing; those who succumb to the fallacy of social continuity envision little or nothing changing as the result of technological innovation. They foresee new technologies doing only old tasks. This habit of mind, an unavoidable product of extrapolating past experience into the future, is demonstrated in a number of the chapters. As Carolyn Marvin has shown, many late-nineteenth-century prophets reacted to Edison's invention of the incandescent bulb in 1879 by imagining a future in which such bulbs would be used for public and theatrical spectacles—precisely the kind of activities for which the existing electric lighting technology, the carbon arc lamp, had been successfully used. So too did the builders of the first digital computers look backward as they made predictions about the future of their new invention.

Having used the computer as a "super-calculator" to solve complex scientific equations, the scientist-inventors mentioned by Ceruzzi expected the machine to be used for similar purposes in the future. They could not imagine computers in wholly different social contexts and functions, such as business use, word processing, or graphic design.

The fallacy of the technological fix shares elements with the other two fallacies. Like the fallacy of total revolution, the idea of the technological fix suggests that a new technology will bring about major changes. But like the fallacy of social continuity, these changes are always envisioned as the resurrecting or strengthening of old social patterns or values, not the introduction of new ones.[3] From the "home of tomorrow," heralded by some prophets as a remedy for the declining American family, to the atomic bomb, which at least one prophet envisioned improving the human species through genetic mutation, a wide range of technologies have appeared as panaceas in the popular imagination. Yet not every invention has prompted such a response. Certain cultural prerequisites are necessary for a technology to be considered a technological fix.

When a technology makes it possible to do something previously thought impossible, particularly if a supernatural aura has surrounded the activity, the new invention becomes a prime candidate for being viewed as a technological fix. The airplane is a classic example, as I have shown in my book *The Winged Gospel.* For millennia people had deemed flight impossible. Only birds or gods could fly. Over the centuries, religious ideas linked flying to omnipotent power and heavenly bliss. Once the Wright brothers had flown at Kitty Hawk in 1903, these associations were projected onto the new flying machine. Flying seemed miraculous, aviation a holy cause, and the airplane a veritable messiah. Prophets glimpsed a dawning air age in which planes would conquer distance, abolish national boundaries, and make all men brothers. They spoke of the airplane as guaranteeing equality, democracy, and perpetual peace—as a kind of universal or omnipotent technological fix.[4]

Electric lighting aroused similar utopian expectations. For thousands of years it was thought that only God could have invented light and rescued humanity from the dread and evil of darkness. The account of creation in Genesis illustrates that thinking: "And God said, Let there be light: and there was light. And God saw the light, that it was good." The flickering light of fires and even the introduction of whale-oil and kerosene lamps did little to foster visions of a technological fix, but with the incandescent bulb mere mortals could flick a switch and, in effect, say "Let there be light." It is not surprising that prophets invested this technology with messianic promise and greeted it as a force for universal betterment. It is likely that the "new light" of x rays, discussed by Nancy Knight, similarly gained credibility as a technological fix because it trespassed on divine territory. Whereas before Roentgen's discovery only God could know the inner secrets of a living creature, now enthusiastic physicians were looking through human tissue and predicting that omnipotent curing power lay just around the corner.

But divine or mystical associations explain only a small portion of the utopian expectations with which Americans have greeted machines. Other technologies, without the kind of cultural associations that influenced popular reception of the electric light or the airplane, also took on the promise of a technological fix. Even so prosaic a material as plastic, scorned in its early days, could become a symbol of utopian hope. How could this happen?

Ignorance about the rudimentary workings of technology, the lack of what we now call technical literacy, has always contributed to the envisioning of material things as social panaceas. Such misunderstanding has become more widespread in the last hundred years as a greater number of new inventions have resulted from the application of abstract and difficult scientific knowledge. Even with the vast expansion of education, complex inventions based on theoretical advances in physics, such as the x-ray machine and the nuclear reactor, were not susceptible to popular understanding. Few fathomed their underlying scientific principles, their performance potentials, or their limitations. Ignorance fostered exaggerated expectations and made possible a magical view of the promise of new inventions. Indeed, it seems that the more closely a technology derived from science, the more readily it prompted utopian hopes. It was the promotion of plastic as a product of modern chemistry, as a result of reactions incomprehensible to the average person, that helped make it seem like an agent of utopian promise. Muddled understanding, then, has often led to visions of a technological fix.

A word must be said about the social ills that prophets have expected new technologies to fix. In general, Americans have imagined that machines would enhance widely shared values, such as democracy, individualism, efficiency, cleanliness, or family stability.[5] Turn-of-the-century predictions that the wireless would allow distant family members to talk regularly with one another, for instance, reflected more than merely a desire for a telephone with a range of over a hundred or so miles. At the time, the family seemed to be reeling from the effects of easier divorce laws, the entry of women into the work force, immigration, contraception, and increased mobility. Anxious speculation over the future of the family filled the popular press. Given this setting, and the American penchant for viewing machines as social panaceas, it seemed plausible that the wireless would provide an electronic substitute for face-to-face intimacy and thereby salvage the family from further decline. No machine proved up to the latter task, of course, yet even today one can hear the same refrain being sung by prophets of the personal computer.

Another example of how visions of the future are constructed out of present concerns and values can be seen in the response to nuclear power in the late 1940s and the 1950s. Claims that the atom would soon make electricity too cheap to meter reflected more than the simple desire to get something for nothing; they owed much to memories of the Great Depression of the 1930s and to the frustrations experienced in the wake of World War II. Amid the Cold War tensions that emerged between the United States and the Soviet Union, made worse when the Soviets detonated an atomic bomb in 1949, the vision of free electricity generated by a peaceful atom served a number of ideological and psychological functions. A prospective nuclear power industry symbolized continued American technical leadership and national prestige, especially since the Soviets themselves were known to be working on nuclear power generation. The dream of American nuclear-generated electricity also helped banish fears, lingering since the Depression, that capitalist institutions could not sustain a high standard of living. With nearly every home wired for electricity and equipped with a burgeoning battery of appliances, prophecies of cheap power implied the continued vigor and superiority of the American way of life.

The second major issue raised by the essays leads one to ask: Just who believed in a future where electricity would be free, where atomic energy would lift airplanes, excavate canals and harbors, or control the weather? Who embraced the buoyant and utopian visions of the future discussed in this book? The essays imply that most Americans shared the prophets' optimism and readily accepted their predictions. Yet some evidence exists, both in the historical literature generally and in the essays here, of a strain of skepticism toward the more sweeping predictions we have been examining.

Brian Horrigan reminds us in his chapter that Lewis Mumford did not share in the euphoria that surrounded discussions of the future of prefabricated housing in the late 1920s and the 1930s. Mumford argued that true reform would come only with significant economic change. In the 1940s and the 1950s, as Steven Del Sesto has told us, a number of scientists disdainfully rejected the more ebullient prophecies made about the atomic future. But we know far less about these naysayers than we should; there is work yet to be done on those individuals who expressed such criticisms and the extent of their influence.

Our understanding of those who spoke optimistically about the future is more extensive. Among the most sanguine regarding the promise of technology were writers for *Popular Science*, *Mechanix Illustrated*, and *Science and Invention*. Popular-science journalism first emerged as a professional specialty at the end of the nineteenth century, when rapid technological innovation, spurred by industrial capitalism, was transforming the economy and everyday life. Popular-science writers not only reported these developments but also predicted the technologies of the future and their expected impact. Whether the subject was x rays in the 1890s, manufactured homes in the 1920s, or computers in our own era, popular-science writers—in Sunday newspaper supplements, popular periodicals, and book-length potboilers—have consistently staked out the high ground of technological utopianism. Indeed, until quite recently, to be popular at all when writing about science and technology almost required that one be upbeat and prophetic.[6]

Many scientists and technologists have been much more cautious and skeptical about the promise of technology. The men who built the first digital computers, John W. Mauchly, J. Presper Eckert, Jr., John von Neumann, and the others, as described by Paul Ceruzzi in his chapter, were typical. Their restrained and almost bleak outlook for the expansion of the computer field derived in part from an intimate understanding of the workings of the new technology as well as from their socialization as physicists and mathematicians. No doubt the desire to avoid embarrassment among their peers also served as a brake on extravagant predictions. In comparison with popular-science writers, scientists and other technically sophisticated insiders to the enterprise of research and development have had fewer incentives to engage in sensationalism and hyperbole. Their training, their work experience, and their professional culture all have tended to dispose them toward more restrained and less utopian expectations for the future. To be sure, there have been exceptions; not all scientists and technologists have been realists about the future, and some have been exuberant utopians. At the moment, however, we know less than we should about the shadings of opinion that existed among scientists and others involved with new technologies and about the factors that shaped their expectations for the future.

We know even less, however, about the views toward technology and the future of those people who did not publish or express publicly their opinions. Women are a case in point. Technological prediction has historically been a predominantly male exercise. This reflects in part the fact that, until very recently, men have not only designed the machines but also built, used, fixed, and even written about them. Specialized masculine cultures grew up around each technology, like the one Susan Douglas describes. These masculine-technical cultures have been little studied, though they are central to technological innovation, adoption, and prediction.[7] These cultures resist penetration by women, not only because of their all-male social composition but also because of their ideology. Recent feminist scholarship has pointed out that the rhetoric of technological mastery and conquest characteristic of these cultures has been stereotypically masculine and thus uncomfortable to many women.[8] Furthermore, the gender-bound nature of this rhetoric, and of the ideology surrounding technology generally, has historically worked to ensure that the future would be predicted by and would be controlled by men.

We must not conclude from this male dominance that women have historically had no thoughts about the future. It can be assumed that every boy wireless enthusiast had a mother and that many of those women were anything but ignorant of the workings and operations of radio. They must have had expectations—whether hopes or fears we do not know—regarding the ways the new technology would shape their future and that of their families. Likewise, women must have had responses to the household technology that, in the 1940s and the 1950s, was packaged by men into so-called kitchens of tomorrow. But what women thought about these developments, or those in other technical areas, is not easily gleaned from the popular-science press, which has rarely published articles about the future written by or about women.

In the same vein, our understanding of how blue-collar workers, farmers, blacks, immigrants, and many other social groups historically conceived the relationship between technology and the future is not likely to be furthered by the study of mainstream literary evidence. Other sources must be consulted, because such groups lacked access to the mass media. In addition, they lacked access to the kinds of education and experience, particularly as inculcated by the various technical cultures, that supported public discourse about the future.

It is often assumed that members of socially disenfranchised groups were consumers of the buoyant visions examined in this book, although good evidence of the precise historical impact of the mass media is hard to find. Indeed the word *impact* may oversimplify the process whereby the mass media influence people. Rather than produce specific effects on a passive audience, the media are part of an interactive process. In the words of three sociologists of communications, "media content, individual needs, perceptions, roles, and values and the social context in which a person is situated" all influence what a person gets out of any particular media offering.[9] If this interactive model is valid, people encountering stories about the coming age of the atom in the 1940s or the 1950s may well have found predictions about future nuclear developments simultaneously entertaining, absurd, and even confirming of their strongest values. Some individuals may have thought such predictions wholly absurd, while others may have greeted them as simply entertaining. From our vantage in time, it would be hard to sort out who felt what. Without the questionnaires and ethnographic techniques of the communications researcher, the historian can only dimly glimpse the popular responses to yesterday's media predictions.

The third major issue to be considered in the light of the above chapters concerns the historical impact of yesterday's predictions about the future. Such visions say much about their period of origins; however, they also have future consequences. We are all familiar with the notion of self-fulfilling prophecies, but prophecies can also influence the future even if they do not accurately anticipate or describe it. Socially, economically, culturally, and even technologically, expectations about the future have contributed to historical change.

Particular visions of the future have facilitated the emergence and the acceptance of new professions. In the 1920s and the 1930s, as Carol Willis suggests, the appeal of the image of the rationalized skyscraper city brought a new degree of fame and recognition to architects and city planners. At the time, many Americans opposed city planning, equating any centralized planning by governments with totalitarianism. Visions of clean, ordered, and efficient skyscraper cities helped undercut such opposition and bring about greater support for planning. In the same period, industrial designers achieved eminence by articulating a streamlined, futuristic vision of the material environment. It is likely that other groups have risen to prominence on the strength of their particular visions of the future, but how common this has been remains a question for further investigation.

Ideas about the future have also had economic consequences. Predictions about tomorrow's technology have been entwined with the development of industrial capitalism—in particular, with the emergence of a society based on mass consumption. Belief in a future transformed by ever-changing products has been an important ideological prop to this consumerism. Companies have exploited this connection in many ways. Most advertising has encouraged prospective fantasy by inviting consumers to imagine how their lives would be altered if they were to acquire the advertised goods or services. Starting in the 1930s, companies began to depict the more distant future, not only in advertisements but also through promotional activities and product design. At the world's fairs of the 1930s, as Folke Kihlstedt and Brian Horrigan have shown, manufacturers "delivered the future" in exhibits and model homes. Such materialized visions of the future bred discontent with existing kitchens, homes, automobiles, and cities. By whetting appetites for an unattainable future, for the purchasable goods of tomorrow, companies sought to ensure that the cycle of consumption—and profit—would continue.

Three broader hypotheses regarding the historical consequences of technological futurism arise from the chapters.

As an ideology, a powerful system of rhetoric and belief, technological futurism seems to have functioned in American culture over the last hundred years much in the same way that Karl Marx saw religion functioning in Europe at an earlier time: as an opiate of the masses. In this formulation, expectations of a halcyon future brought forth by technology have had an anesthetic effect. The ability to deal with the social and political dimensions of a problem has been dulled by the euphoric solutions proferred by technological utopians. Faith in technology (or, more accurate, in the future promised by technology) became not only a kind of secular religion but also a substitute for politics. The optimism with which some Americans in the 1920s and the 1930s looked forward to the home of tomorrow is a typical instance. Hopes for better housing reposed in a sleek glass-and-steel technological fix which by definition could not redress aspects of the housing crisis arising from mortgage practice, land speculation, restrictive covenants and zoning, or inadequate income. To the extent that people believed in the myth of a future transformed and made better by things, support for real social change was undermined.

Popular futurism has also been an important component of nationalism in modern American history. Beginning in the late nineteenth century, the future appears to have played a role similar to that of the past in an earlier stage of American development. Much as various intellectuals in the first half of the nineteenth century invented a mythic past for the new nation, providing substitute myths for the ancient traditions that shaped European national identities,[10] popular futurists in the last hundred years have mythologized the future, inventing an America that is harmonious and affluent because of science and technology. This vision of the technological future has resonated in American culture precisely because the actual technological changes of the last century have been accompanied by so much privation, conflict, and dislocation. The future became a fun-house mirror image of what existed in the present. Tomorrow's technology would be a colossal fix for the wrenching transformations brought about by industrial capitalism.

Finally, utopian expectations may paradoxically have helped bring about workable innovations. The most naive fantasies about machines and their possible impact are part of the same cultural milieu in which actual invention takes place and technology is adopted and diffused. Although scholars usually explain the activity of inventors as a response to market demand, it is plausible that inventors have responded as well to popular dreams and expectations.[11] The chapters provide little direct support for this hypothesis; however, in a culture that celebrates progress and invests machines with utopian promise, it is possible that inventors have been motivated by values other than the acquisition of wealth and fame. I would propose that some inventors, at least, have resembled less the rational and purposeful economic man of scholarly theory than what we might call the visionary man, a kind of dreamer who imagined that his mechanical contrivances could solve social problems.

Although the authors of this volume tend to be skeptical of that belief, technological utopianism remains a potent force in American culture. Faith in a better tomorrow through technology continues to influence policy and events. As we anxiously ponder our destiny on the eve of the twenty-first century, striving to understand and manage technologies of fearful complexity and danger, an understanding of how people in the past thought about technology and the future may provide useful insights and even a modicum of humility.

Notes

1. For publications relevant to this historical study of thinking about the future, see the notes for the introduction to this volume (especially note 5).

2. See Nathan Rosenberg, "Factors Affecting the Diffusion of Technology," in N. Rosenberg (ed.), *Perspectives on Technology* (Cambridge University Press, 1976), p. 193; N. Rosenberg, "Learning by Using," in N. Rosenberg (ed.), *Inside the Black Box: Technology and Economics* (Cambridge University Press, 1982), pp. 120 ff.

3. There is no doubt that the introduction of certain technologies (or their removal, as in the example of handguns) can decidedly ameliorate particular social problems.

For a discussion of such technological fixes, see Amitai Etzioni and Richard Remp, *Technological Shortcuts to Social Change* (New York: Russell Sage Foundation, 1973).

4. Joseph J. Corn, *The Winged Gospel: America's Romance with Aviation, 1900–1950* (New York: Oxford University Press, 1983).

5. The American tendency to view technology through the prism of democratic values is well developed in John Kasson's *Civilizing the Machine: Technology and Republican Values in America, 1776–1900* (New York: Viking, 1976).

6. *Popular Science* and *Popular Mechanics* remain bastions of technological utopianism, unaffected by the more cautious movement of technology assessment of recent years. The computer field, too, has generated an immense outpouring of utopian predictions, as books with titles such as *The Micromillennium* attest. For critiques of the messianic faith in computing, see Carolyn Marvin, "Fables for the Information Age: The Fisherman's Wishes," *Illinois Issues Humanities Essays* 17, second series (September 1982), pp. 17–24; Langdon Winner, "Mythinformation in the High-Tech Era," *IEEE Spectrum*, June 1984, pp. 90–96.

7. For a recent study in which ethnographic techniques are used to document and analyze the cultures of computer "hackers" and artificial-intelligence workers, see Sherry Turkle, *The Second Self: Computers and the Human Spirit* (New York: Simon and Schuster, 1984).

8. See, for example, Carolyn Merchant, "Mining the Earth's Womb," in Joan Rothschild (ed.), *Machina Ex Dea* (New York: Pergamon, 1983).

9. Dennis McQuail, Jerry G. Blumler, and J. R. Brown, "The Television Audience: A Revised Perspective," in D. McQuail (ed.), *Sociology of Mass Communications: Selected Readings* (Harmondsworth: Penguin, 1972), p. 144.

10. See the title essay in Henry Steele Commager's *The Search for a Usable Past and Other Essays in Historiography* (New York: Knopf, 1967).

11. An excellent review of the scholarly literature dealing with technical innovation is Nathan Rosenberg's "The Historiography of Technical Progress," in his *Inside the Black Box*.

About the Authors

Paul Ceruzzi is Associate Curator for Space Science and Exploration at the National Air and Space Museum in Washington, D.C., and a research associate of the Computer Museum in Boston. He is the author of *Reckoners: The Prehistory of the Digital Computer, from Relays to the Stored Program Concept, 1935–1945* (Westport, Conn.: Greenwood, 1983).

Joseph J. Corn teaches in the American Studies Program and the Program in Values, Technology, Science, and Society at Stanford University. He is the author of *The Winged Gospel: America's Romance with Aviation*, and a co-author (with Brian Horrigan) of *Yesterday's Tomorrows: Past Visions of the American Future* (New York: Summit, 1984).

Steven L. Del Sesto is a sociologist who has published widely on historical and political aspects of nuclear energy. He is currently employed by Kidder, Peabody & Smith in Providence, Rhode Island.

Susan Douglas is Assistant Professor of Media and American Studies at Hampshire College. She is finishing a book on the early history of radio.

Brian Horrigan works as a consultant and writer on topics relating to history, technology, and culture. He served as guest curator (with Joseph Corn) at the Smithsonian Institution for the exhibit "Yesterday's Tomorrows: Past Visions of the American Future" and co-authored the exhibit book of the same name.

Folke T. Kihlstedt is Associate Professor in the Deparment of Art at Franklin and Marshall College. His main interest is in the history of building techniques and architectural studies, and he is working on a book to be entitled *The Wheels of Modernism.*

Nancy Knight is Museum Specialist in Medical Sciences at the National Museum of American History. She is the author of *Pain and Its Relief* (1983) and is currently directing the preparation of a museum of modern health technology near Fort Lauderdale, Florida.

Carolyn Marvin is Assistant Professor at the Annenberg School of Communications at the University of Pennsylvania. She is the author of a forthcoming study on the introduction of new electric communications technologies in the late nineteenth century.

Jeffrey L. Meikle is Assistant Professor of American Studies and Art History at the University of Texas at Austin. He is the author of *Twentieth Century Limited: Industrial Design in America, 1925–1939* (Philadelphia: Temple University Press, 1979).

Howard P. Segal is a visiting lecturer in the History of Science department at Harvard University. He is the author of *Technological Utopianism in American Culture* (University of Chicago Press, 1985).

Carol Willis is finishing her dissertation on American visionary architecture of the 1920s and the 1930s in the Department of Art History and Archeology at Columbia University. She is curator for the exhibit "Hugh Ferris: Metropolis" for the Institute for Architecture and Urban Studies.

Index